SpringerBriefs in Applied Sciences and Technology

Automotive Engineering: Simulation and Validation Methods

Series editors

Anton Fuchs, Graz, Austria
Hermann Steffan, Graz, Austria
Jost Bernasch, Graz, Austria
Daniel Watzenig, Graz, Austria

For further volumes:
http://www.springer.com/series/11667

Alexander Thaler · Daniel Watzenig

Editors

Automotive Battery Technology

Editors
Alexander Thaler
Daniel Watzenig
Virtual Vehicle Research Center
Graz
Austria

ISSN 2191-530X   ISSN 2191-5318 (electronic)
ISBN 978-3-319-02522-3 ISBN 978-3-319-02523-0   (eBook)
DOI 10.1007/978-3-319-02523-0
Springer Cham Heidelberg New York Dordrecht London

Library of Congress Control Number: 2013958389

Printed on acid-free paper

Springer is part of Springer Science+Business Media (www.springer.com)

Foreword

Ongoing discussions about climate change and the related fuel economy challenges are moving the automotive industry more and more in the direction of pure electric driving. At this point, it is no longer a question of "if" electromobility will become reality, but rather a question of "when" it will become a reality, and which applications will be first to market. Worldwide legal reductions in carbon dioxide and emissions limits will essentially require the electrification of automobiles. In addition, there is an ongoing shift in consumer attitudes towards electric vehicles, which is evident in two examples: Tesla's electric Model S has already been very successful, and BMW plans to launch its electric i3 by the end of 2013. Those two models, which feature completely new vehicle technology that has been specifically developed for an electric powertrain, may already herald a new era of individual mobility.

One of the main factors driving this trend is that battery technology, which is at the heart of electromobility, has improved significantly in recent years. Particularly in the field of lithium-ion batteries, which are essential for plug-in vehicles (PHEV) and fully electrified vehicles (BEV) due to their high energy and power density, significant progress has been made in reducing costs and improving safety, performance and reliability. For example, energy density has been increased significantly, without compromising power capability. Today's 18650 lithium-ion cells, which are already in use in automotive applications, can achieve a capacity of more than 3Ah. Safety has also been increased significantly, such as via new separator technologies and/or improved chemistries (e.g. $LiFePO4$). Indeed, many measures have already been implemented in the battery management system to protect the battery from dangerous external events (e.g. overcharge, over-temperature, over-current).

However, from a consumer's perception point of view in automotive applications and over 20 years of experience in other commercial applications (e.g. mobile phones) have shown that there is still considerable room for improvement if drivers are to fully embrace electric driving and the related battery technology. Costs need to be further reduced, and the reliability, durability and safety of lithium-ion batteries can be enhanced. In terms of safety, for example, more must be learned about system safety and abuse tolerance under vehicle crash conditions. Such advances will require an even deeper understanding of the microscopic processes in batteries under both normal and abnormal conditions (e.g. misuse or crash situations). Furthermore, new analytical approaches must be developed to provide an enhanced insight into the electrochemical processes within the lithium-ion battery cells, which will in turn improve the accuracy of estimations of the remaining vehicle driving range by enabling a better state-of-charge determination of the cells.

This book contributes to this ongoing development effort by providing insight into the state of the art of lithium-ion battery technology modelling. It is designed to stimulate new ideas for the improved utilization of batteries in real application. From a long-term perspective, we hope that this book will help foster improvements of the technology itself and thereby help usher vehicle technology into the next era. While today's lithium-ion batteries are suitable for application in electric cars, it is clear that many improvements are still to come in terms of both the physics and chemistry of such batteries. Energy and power density, which directly affect driving range and vehicle performance, will be significantly increased; safety and reliability will be further improved; and the total cost of batteries will decrease once production figures reach automotive scale quantities. As a consequence, consumer acceptance will increase significantly, as vehicles driven by "horsepower" from combustion engines will become less attractive than quiet, environmentally friendly electric vehicles propelled by "kiloWatts". When this happens, PHEVs and BEVs will have truly arrived.

Graz, November 2013   Volker Hennige

Preface

During the last 10 years, the modern world has seen at least the second attempt to electrify the powertrains of road traffic vehicles. Beyond the battery, which is a key component in this electrification process, all other parts have been further developed. All power devices have made a huge step forward. The combination of powerful control devices and semiconductor enhancements are providing adequate functionality from a customer point of view.

Much research has been conducted on the electrochemical level, and there has been progress made on energy density and cost as well. However, currently there is no electrochemical energy storage system available that fulfils the energy demand of today's drive trains and the related passenger comfort functions. Of course, it is simple to blame the battery and argue that this technology does not cover the needs. On the other hand, we can see that society is embracing the need to be more efficient when using energy, a need which is particularly strong in the area of mobility. Thus, the trend towards the electrification of drive trains is revealing the essential weakness of today's vehicle concepts. In the past, the availability of fossil fuels, with their high energy density, fostered vehicle evolution. Due to the limitation of on-board installed energy in storage systems, public awareness of the low efficiency of current vehicles has been growing, and the automotive industry has grasped the demand for more efficiency.

Two developments are anticipated:

- An increase in efficiency on the drive-train level, comfort (HVAC, etc.) and safety functions within a vehicle
- An increase in energy density in terms of energy storage:

    - Energy density will increase on the chemical level
    - Technology integration aspects (e.g. ageing, safety) will become better understood and development processes will be implemented in the automotive industry.

This book provides a comprehensive overview of the current research work in the latter area (i.e. integration). The two main topics of this book, safety and ageing, both directly influence the size and utilization of the applied storage system.

Articles Related to Safety

In Chap. 1, Martin et al. offer an overview of today state-of-the-art safety standard. Although this standard is well defined in ISO 26262, from an overall system-safety perspective, important processes and methods are still missing. Since safety aspects influence cost, it is essential to understand how different safety measures reduce risk for a new product (e.g. a battery system) that is integrated into the vehicle environment.

In Chap. 2, Trattnig and Leitgeb provide an overview of the challenge of mechanical battery modelling in crash/crush battery simulations. The current challenge in this area is to bridge the gap between the battery's micro structures and the need to keep simulation effort manageable. The question is, how simple can a model be made while still preserving its ability to provide all of the necessary information at the vehicle level to enable crash-relevant optimizations.

In Chap. 3, Golubkov and Fuchs focus on the thermal runaway process. Their team is currently working to develop a basic, application-related understanding of this process. The knowledge of this process will enable the creation of a battery system simulation framework that can predict the propagation of thermal runaway within the whole battery and the vehicle as a whole.

Articles Related to Ageing

In Chap. 4, Pichler and Cifrain describe an approach for modelling the electro-chemical battery cell with all of the necessary details. The major challenge here is to devise a model that covers the physical properties of the cell in nanometre scale (e.g. anode/cathode porosity) while still providing simulation output in a reasonable amount of time and with an acceptable level of quality. The final step is the optimization of the cell design and technology in front of the application (e.g. driving cycle), while also covering major ageing aspects. A detailed model of physical processes always requires real parameters derived directly from physical measurements. In Chap. 5, Weber et al. provide an overview of analytical methods for quantifying the ageing of lithium-ion batteries. Such laboratory work is a necessary input for the models mentioned above. Since a complex model uses parameters that are not directly measureable, in Chap. 6, Scharrer et al. present a mathematical method for parameter optimization. To demonstrate this method, they present the results of a synthetic fitting problem solved by a parallel-adaptive Markov chain Monte Carlo method.

In Chap. 7, Hametner and Jakubek present a data-based, chemistry-independent approach to nonlinear observer design for the state-of-charge (SoC) estimation. In order to operate the energy storage system throughout the required lifetime, knowledge of the SoC is essential, and one of the key factors related to ageing.

One significant challenge for all of these individual approaches towards improved safety and life time of battery systems is the complexity this component adds to the car. Standards in the automotive industry are particularly high, especially in terms of quality and durability, and all research conducted in these fields must be measured against these standards. In the field of applied research, the key to meeting these high standards is the combination of knowledge from different specialist domains.

Such collaboration can produce high quality, highly useful development environments (modelling, simulation tools, accompanying tests and standards). In this context, the coordination of the efforts of specialists from different industries and from research institutes is the way forward.

Graz, November 2013   Alexander Thaler
                                                                       Daniel Watzenig

Contents

Contributors

Martin Cifrain Virtual Vehicle Research Center, Inffeldgasse 21a, 8010 Graz, Austria, e-mail: martin.cifrain@v2c2.at

David Fuchs Virtual Vehicle Research Center, Inffeldgasse 21a, 8010 Graz, Austria, e-mail: david.fuchs@v2c2.at

Andrey W. Golubkov Virtual Vehicle Research Center, Inffeldgasse 21a, 8010 Graz, Austria, e-mail: andrej.golubkov@v2c2.at

Heikki Haario Lappeenranta University of Technology, Lappeenranta, Finland, e-mail: heikki.haario@lut.fi

Christoph Hametner Christian Doppler Laboratory for Model Based Calibration Methodologies, Vienna University of Technology, Wiedner Hauptstr. 8-10, 1040 Vienna, Austria, e-mail: christoph.hametner@tuwien.ac.at

Stefan Jakubek Institute of Mechanics and Mechatronics, Vienna University of Technology, Wiedner Hauptstr. 8-10, 1040 Vienna, Austria, e-mail: stefan.jakubek@tuwien.ac.at

Werner Leitgeb Virtual Vehicle Research Center, Inffeldgasse 21a, 8010 Graz, Austria, e-mail: werner.leitgeb@v2c2.at

Andrea Leitner Virtual Vehicle Research Center, Inffeldgasse 21a, 8010 Graz, Austria, e-mail: andrea.leitner@v2c2.at

Helmut Martin Virtual Vehicle Research Center, Inffeldgasse 21a, 8010 Graz, Austria, e-mail: helmut.martin@v2c2.at

Sascha Nowak MEET Battery Research Center, Westfälische Wilhelms-Universität Münster, Corrensstr. 46, 48149 Münster, Germany, e-mail: sascha.nowak@uni-muenster.de

Franz Pichler Virtual Vehicle Research Center, Inffeldgasse 21a, 8010 Graz, Austria, e-mail: franz.pichler@v2c2.at

Falko Schappacher MEET Battery Research Center, Westfälische Wilhelms-Universität Münster, Corrensstr. 46, 48149 Münster, Germany, e-mail: falko.schappacher@uni-muenster.de

Matthias K. Scharrer Virtual Vehicle Research Center, Inffeldgasse 21a, 8010 Graz, Austria, e-mail: matthias.scharrer@v2c2.at

Alexander Thaler Virtual Vehicle Research Center, Inffeldgasse 21a, 8010 Graz, Austria, e-mail: alex.thaler@v2c2.at

Gernot Trattnig Virtual Vehicle Research Center, Inffeldgasse 21a, 8010 Graz, Austria, e-mail: gernot.trattnig@v2c2.at

Daniel Watzenig Virtual Vehicle Research Center, Inffeldgasse 21a, 8010 Graz, Austria, e-mail: daniel.watzenig@v2c2.at

Sascha Weber MEET Battery Research Center, Westfälische Wilhelms-Universität Münster, Corrensstr. 46, 48149 Münster, Germany, e-mail: sascha.weber@uni-muenster.de

Bernhard Winkler Virtual Vehicle Research Center, Inffeldgasse 21a, 8010 Graz, Austria, e-mail: bernhard.winkler@v2c2.at

Chapter 1
Holistic Safety Considerations for Automotive Battery Systems

Helmut Martin, Andrea Leitner and Bernhard Winkler

Abstract The objective of system safety engineering is to develop a system with no unreasonable risk. To this end, risks caused by the electrical and/or electronic (E/E) system that could potentially harm persons must be analyzed, and appropriate risk reduction measures have to be considered in an early phase of development. This requires a close collaboration between different engineering disciplines in order to specify a comprehensive description of risk reduction and mitigation measures—the safety concept. The international functional safety standard ISO 26262 has to be considered for the development of E/E systems within road vehicles up to 3.5 tons. This standard focuses on E/E measures and considers other non-E/E measures only after the specification of the safety concept. In contrast, this chapter proposes a workflow for the elaboration of an integrated safety concept including safety measures from different engineering disciplines. Two main lessons learned were that the consideration of all kinds of risk reduction measures in the concept phase improves the understanding of the safety of the overall system, and involving various fields of expertise enables the development of a clear safety concept. This approach will improve the development of the overall system, while complying with the requirements of ISO 26262 for the development of E/E systems. The applicability of the introduced approach is demonstrated on an automotive battery case study, where the influence of various safety measures on the Automotive Safety Integrity Level (ASIL) determination has been taken into account in order to reduce the costs of E/E system development.

H. Martin (✉) · A. Leitner · B. Winkler
Virtual Vehicle Research Center, Graz, Austria
e-mail: helmut.martin@v2c2.at

A. Leitner
e-mail: andrea.leitner@v2c2.at

B. Winkler
e-mail: bernhard.winkler@v2c2.at

A. Thaler and D. Watzenig (eds.), *Automotive Battery Technology*,
Automotive Engineering: Simulation and Validation Methods,
DOI: 10.1007/978-3-319-02523-0_1, © The Author(s) 2014

1.1 Motivation

Hazardous voltage (HV) battery systems are a central part of battery-powered Electric Vehicles (EVs) or Hybrid Electric Vehicles (HEVs) [8], which are becoming more and more important. One reason is the high energy efficiency of E/E systems and the zero (local) environmental pollution of EVs. Their main disadvantage is the relatively short operation range, which is far less competitive compared to conventional vehicles with internal combustion engines. Conventional vehicles provide good performance and long operating ranges by utilizing the high energy-density advantages of petroleum fuels. HEVs combine the advantages of both technologies. Some of the main targets for batteries to be used in HEVs are low costs, high power density (e.g. 1,200 W/kg), very high cycle life time (e.g. 200,000 cycles of charge/discharge), long life time (e.g. 9 years), and safety. With the growing importance of e-mobility, automotive battery systems are becoming more important as well. High power (e.g. HEV up to 250 kW to provide more dynamic driving torques) and high energy application (e.g. EVs such as Nissan Leaf 36 kWh to allow longer driving distances) are already being applied in series-production vehicles. Increasing power and energy while decreasing the battery geometries leads to an increase of potential critical effects in the case of malfunctions.

This chapter focuses on safety aspects in the context of safety-critical automotive batteries for EVs or HEVs. Regarding functional safety (safety of the E/E system), the IEC 61508[1] [3] is the basic international functional safety standard applicable to all industries. The ISO 26262 [4] is an adaptation of this standard that is applicable to the development of safety-related electrical and/or electronic (E/E) systems in the automotive domain. One important aspect of functional safety is the potential risk of electronic malfunction, e.g. malfunction of the battery control unit caused by incorrect inputs or software errors. These malfunctions could lead to hazardous events for passengers, other traffic participants, and uninvolved parties (e.g. fire due to overcharge). The potential of malfunctions has to be lowered by gaining of possible faults, as well as their causes and effects, and by providing solutions for fault mitigation.

In particular, e-mobility is highly interdisciplinary, whereby risk reduction also results from different technical disciplines (e.g. mechanics, chemistry). This means that system safety has to consist of different safety disciplines as well (i.e. functional, electrical, mechanical, and chemical safety). One example for electrical safety could be the prevention of hazardous voltage through the use of galvanic disconnections or isolation. Mechanical safety aims to prevent the deformation of the battery in the case of an accident through the use of cell housings or the installation location for example. Chemical safety can prevent explosions or fire by using a mechanical venting outlet for toxic gases. All of these measures are applicable for the development of a safe system.

---

[1] IEC 61508—Functional safety of electrical/electronic/programmable electronic safety-related systems.

Functional safety covers one vital part of system safety engineering, but it is important to realize that other safety measures have to be considered as well. This chapter discusses some of the main issues regarding the safety of HV automotive battery systems on different levels of abstraction such as battery cell, battery module and battery pack.

This chapter is structured as follows: Sect. 1.2.1 starts with an introduction to the safety lifecycle following ISO 26262. Section 1.2.2 describes the technical background consisting of the basic architecture of a battery system, together with potential risks and risk mitigation on different levels of abstraction. To get a better understanding, these safety measures are classified in Sect. 1.3. Section 1.4 introduces a modified workflow, which is used to reduce the required Automotive Safety Integrity Level (ASIL) and thereby also the development costs of the electronic system through the definition of non-E/E measures. Section 1.5 concludes the work and provides an outlook on how the presented work will be continued.

1.2 Technical Background

This section introduces the topic of functional safety in the context of automotive systems. Furthermore, an overview of an HV battery system architecture is provided, including several basic safety measures from different engineering disciplines.

1.2.1 Introduction to Functional Safety Following ISO 26262

The ISO 26262 safety lifecycle encompasses the principal safety activities during the concept phase, product development, production, operation, service and decommissioning as illustrated in Fig. 1.1.

Figure 1.1 shows the safety lifecycle and highlights the concept phase and the relevant parts of the product development. The concept phase starts with the definition of the system (here called item), followed by a Hazard Analysis and Risk Assessment (HA&RA), in which all identified hazardous events are evaluated according to ISO 26262 specific risk assessment criteria (i.e. severity, exposure and controllability). Current hazard analysis techniques can be classified on a hierarchical structure of a system in bottom-up (e.g. FMEA) and top-down approaches (e.g. FTA). The most important, often-cited techniques for performing a hazard analysis are Preliminary Hazard Analysis [1, 6], Concept Failure Mode and Effects Analysis (Con-FMEA) [2], and Hazard and Operability study (HAZOP) [5]. By performing the hazard analysis we identified the following main hazards of the battery system: fire/explosion, toxic gases, hazardous voltage of the battery module/pack (U > 60VDC), leakage/venting of battery cells (corrosive/toxic (e.g. hydrofluoric acid)), fire (e.g. flammable materials) and explosion (e.g. breakdown of cell safety vent).

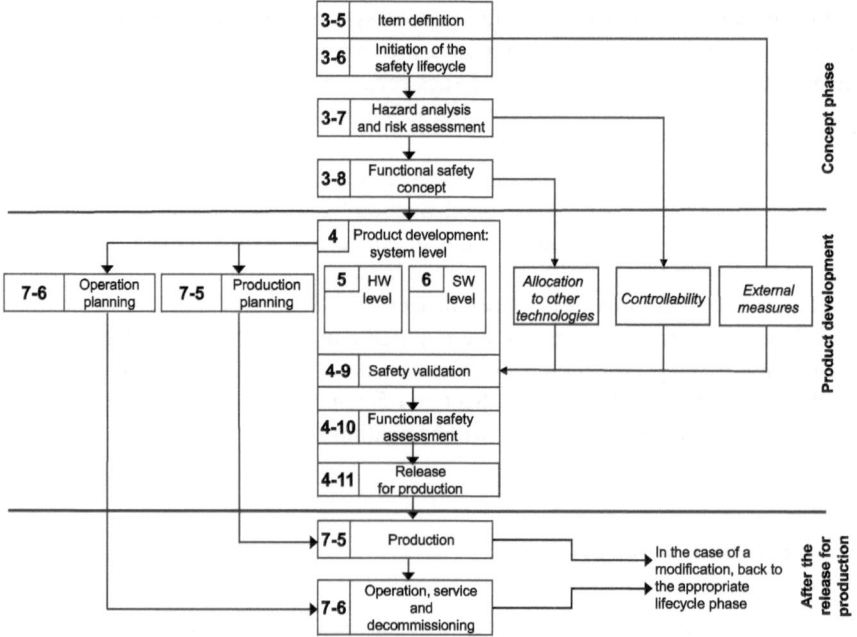

Fig. 1.1 Safety lifecycle according ISO 26262 [4]

The result of the risk assessment determines the ASIL, which indicates the risk of occurrence of a specific failure mode[2] and its necessary degree of avoidance. ASIL values range from ASIL A (low criticality) to ASIL D (high criticality).[3] Depending on the derived ASIL, the ISO 26262 recommends methods for fulfilling the requirements—higher ASIL leads to higher efforts and costs during the product development.

Based on the results of the HA&RA, safety goals[4] are defined for each hazardous event, and the corresponding ASIL is allocated to each of them. The final activity of the concept phase is the elaboration of the Functional Safety Concept, which defines safety measures that must be fulfilled by the design and development of the system to avoid an unreasonable residual risk. Safety measures are activities or technical solutions used to avoid, control or mitigate the harmful effects of systematic failures and random hardware failures. These technical solutions are implemented by (i) E/E measures (e.g. E/E system with sensor → controller → actuator), (ii) external measures (e.g. organizational measures to counter technical flaws) or (iii) other technologies (solutions from other technical domains, e.g. mechanical

---

[2] *"failure mode = manner in which an element or an item fails"*. [4]

[3] The class QM (quality management) denotes no requirement to comply with ISO 26262.

[4] Safety goals represent top level safety requirements.

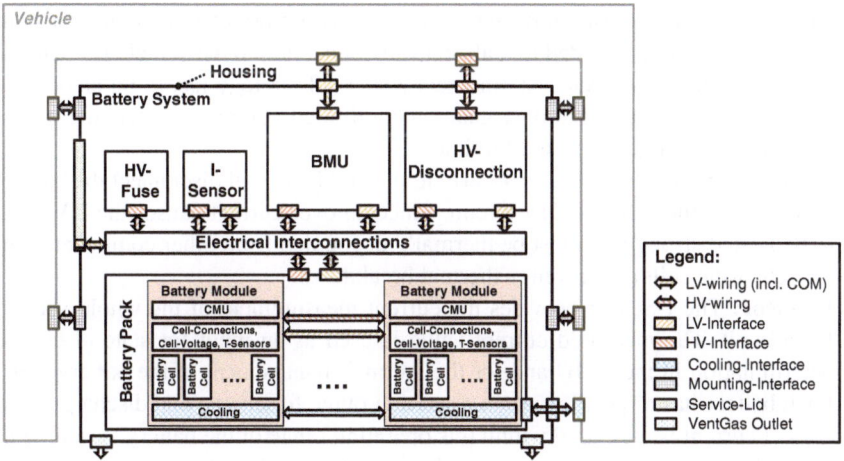

Fig. 1.2 Battery system architecture—Illustration of the main parts of an automotive battery and their interrelations

fault-back solution), which detect faults or control failure modes in order to achieve or maintain a safe state.[5]

1.2.2 Description of Automotive Battery System Architecture

Figure 1.2 shows a schematic representation of a system architecture of an HV lithium-ion battery. It consists of the following main components, which already include or represent basic safety measures:

- **Battery Management Unit (BMU):** The main functions of the BMU are the electrical and thermal management, diagnosis functions, insulation monitoring, and the communication with other parts of the vehicle. Electrical management includes charge balancing, charge determination, and the provision of status information, such as system voltage, system current, or power-time prediction (charging/discharging) for vehicle control functions. Thermal management functionality is used to monitor and evaluate the temperature in the battery system. Disconnection monitoring, charge monitoring, and fault recording represent different diagnosis functions. The insulation monitoring in the battery system is a coordinated function between the battery system and the vehicle.
- **HV Disconnection:** Its main purpose is the disconnection of the battery system from the vehicle HV circuit, and it provides a galvanic separation of the battery and the vehicle in case of deactivation, accident or a safety-critical malfunction. The HV disconnection consists of special HV contactors for the plus and minus

---

[5] *"safe state = operating mode of an item without an unreasonable level of risk of the system".* [4]

terminal. For the activation of the system, a specific pre-charge circuit for both terminals has to be included to realize a soft connection to the vehicle HV circuit. In case of an over-current, an emergency shut-off strategy has to be elaborated because the contactors can only guarantee a limited number of switching cycles under load over their expected lifetime.

- **HV Fuse:** In the case of an over-current, the HV Fuse will disconnect the battery system from the vehicle's HV circuit. Since an over-current causes the HV Fuse to be heated strongly, it must be thermally decoupled from other components (in particular the cells) to prevent a thermal breakdown.
- **I-Sensor:** The I-Sensor provides the current measurement of the whole vehicle HV circuit. The measured current value is used as an input for state-of-charge determination in the BMU and for the thermal management of the battery cells. Each battery has a specific current operation range for charge and discharge. The correct current is measured within this operating range of the battery system with a specified accuracy. If the current is lower or higher than the operating range, a special disconnection strategy has to be implemented with interaction of the HV Disconnection and the HV Fuse.
- **Electrical Interconnections:** This includes all kinds of LV (low voltage wiring including the communication) and HV connections between the battery cell pack and the relevant E/E components of the battery system.
- **Battery Cell Pack:** The battery cell pack consists of serial and/or parallel-connected battery cell modules and the battery cell module interconnection.

  - **Battery Cell Modules** consists of battery cells that are connected in series and/or parallel and a cell management unit (CMU). The CMU is responsible for cell charge balancing, measurement of cell voltage and temperature, and the communication between CMUs in different battery modules as well as between CMU and BMU. The cell modules contain a number of redundant temperature sensors to detect areas with critical temperatures. These sensors are connected with the thermal management in order to prevent critical temperature in the battery system.
  - **Battery Cell Module Interconnection** includes all electrical, mechanical, and thermal connections between battery modules.

- **Housing and external interfaces:** The main purpose of the battery housing is to protect the battery system from environmental influences and to protect the driver from any unintended reaction of the battery system. It prevents people from coming into contact with any hazardous voltage. Furthermore, the housing couples the battery system and the vehicle. It has to provide a LV (including communication), an HV interface and an interface for cooling. The housing should provide vent gas outlets (vent gas management), in case of an overpressure in the battery system. For maintenance and repair of the battery system, a service outlet is available. The mechanical mounting interface connects the battery with the vehicle bodyworks.

1.3 Classification and Application of Safety Measures for Automotive Battery Systems

As mentioned before, it is reasonable to consider different types of measures in order to achieve a more holistic safety concept. Some of these measures are given by customer requirements, while others have to be introduced for additional safety reasons. In this section, we classify them in organizational and technical measures and show some examples.

1.3.1 Organizational and Technical Safety Measures

This work classifies safety measures in two main categories:

- **Organizational safety measures [ORGA]** encompass:

  Safety-compliant development process: The company-specific development process has to cover relevant safety-standard-specific process activities (e.g process audits by external bodies).

  Review/Inspection/Confirmation: Work products that make up the safety case have to be checked by independent[6] parties.

  User safety manuals: Clear and understandable manuals and instructions for the correct handling of the product in the native language of the end user are required.

  Warning labels and signs indicate potentially critical parts of the system that could cause harm to people (e.g. vent gas outlet at battery housing).

  Training: End users have to be informed/trained how to handle the product (e.g. correct driver reaction in the case of malfunction of the battery system). Some kind of safety training is also necessary for first responders in the case of an accident because they should be able to rescue people and should not endanger themselves.

  Transport/Storage Regulations: Test and criteria are defined for transport and storage-specific scenarios that have to be approved for the battery cells (UN/ADR regulations e.g.UN 38.3 [10]).

  Periodicity of maintenance: The proper functioning of the different safety measures has to be guaranteed until the product's decommissioning. Instructions for maintenance, repair and decommissioning of the product are defined in the standards as well.

- **Technical safety measures encompass:**

  Functional safety [FUSA] : Possible malfunctions of the battery system should be avoided, mitigated, or handled by adequate E/E safety measures (e.g.

---

[6] The degree of independence depends on the safety integrity level, which is defined in the concept phase.

detection of overcharge of battery and disconnect the battery from any external energy source). This kind of safety measure is explicitly covered by ISO 26262. In contrast, the following other technical safety measures are referred to as external measures or other technologies.

Chemical [CHEM]: Any kind of reduction of toxicity of chemical substances (e.g. chemical proof material, cell chemistry) and mitigating the effects of any hazardous cell reaction.

Thermal [THER]: Reduction of thermal energy (e.g. cooling of cells).

Electrical [ELEC]: Avoidance of hazardous voltages for customers (e.g. electrical insulation).

Mechanical [MECH]: Mechanical construction should prevent or mitigate harm caused by external source.

1.3.2 Application of Measures at Battery System Units

Not only the incorporation of different engineering disciplines, but also the investigation and coverage of safety at the appropriate level of detail is important for the development and production of a safe system (see Fig. 1.2). This section discusses the different levels of units of an automotive battery system. The investigation starts from the lowest level (i.e. the cell) and ends with the highest level (i.e. the vehicle where the battery should be integrated). The battery system is separated into different units, and examples of safety measures are provided.

Level 4: Battery Cells (BatCel)

This level focuses on all relevant aspects of cell design and structure, cell housing, possible vent gas outlets, cell behavior during ageing over life cycle of the battery, and so on.

Sample safety measures:

- [ORGA] Cell production process—Establishment of battery cell production quality process, to avoid any kind of contamination of the cell during the production process.
- [CHEM] Cell structure—Choice of chemical cell components (e.g. cathode, electrolyte additives).
- [MECH] Charge Interruption Device—Mechanical construction in the cell. It is activated if anything causes internal cell pressure to exceed the activation limit physically, and it will irreversibly disconnect the cell from the circuit.
- [MECH] + [THERM] Thermal management—Cooling and heating of cells, if needed.
- [ORGA] + [MECH] + [CHEM] Vent gas management—Each battery cell provides a defined mechanical venting opening in case of a cell defect.

Level 3: Battery Module (BatMod)

The battery module level covers various safety measures for the different interfaces of the cells to build up a so-called battery cell stack. One argument for the packaging of cells in modules is the fact that modules can be replaced during maintenance.

Sample safety measures:

- [MECH] + [THERM] Use of materials that absorb thermal energy in the module (increase of thermal capacity).
- [MECH] + [THERM] Thermal management—Cooling and heating of the cells if needed
- [ORGA] + [MECH] + [CHEM] Vent gas management—Each battery module provides a defined mechanical venting opening in case of a cell defect.
- [FUSA] Monitoring of cell balancing—If a fault is detected (e.g. overcharge), transition to safe state in that situation.

Level 2: Battery Pack (BatPack)

The battery pack encompasses all modules and provides electrical, thermal, and mechanical connections between them.

Sample safety measures:

- [MECH] + [THERM] Thermal management—Cooling and heating of the cells if needed.
- [ORGA] + [MECH] + [CHEM] Vent gas management—BatPack combines all vent gas channels from each BatMod and leads it to the BatSys.

Level 1: Battery System (BatSys)

The battery system contains the battery pack, the housing, the BMU, and other relevant components. The BMU internally coordinates all parts of the battery and provides an interface to the E/E system at the vehicle level. It is therefore responsible for the detection and mitigation of errors from the external system.

Sample safety measures:

- [ORGA] + [MECH] + [CHEM] + [THERM] Fire extinguisher inlet—The BatSys system should provide an inlet so that the fire brigade could keep the fire at bay and cool down the battery cells.
- [ORGA] + [MECH] + [CHEM] Vent gas management—BatSys provides a vent gas outlet at the battery housing for the vehicle.
- [FUSA] The BMU is an E/E system and is responsible for e.g. monitoring of cell breakdown—If a cell break down is detected by the BMU, several actions should be triggered: disconnection of battery, increase of cooling, communication of critical battery fault.

Level 0: Vehicle Level (target integration of battery system)

At the vehicle level, the prerequisites for the correct functioning of the battery system must be clearly defined. Battery system vendors have to make assumptions about the

Malfunction: Overcharge / Safety Measure	Safety Discipline						Level				
	CHEM	THER	ELEC	MECH	FUSA	ORGA	L4-BatCel	L3-BatMod	L2-BatPack	L1-BatSys	L0-VEH
Vent gas management	x			x		x	x	x	x	x	x
Cell voltage monitoring			x		x		x	x	x	x	
Charge Interruption Device (CID)				x			x				
Monitoring of cell balancing		x	x		x			x		x	
Thermal management		x		x	x		x	x	x	x	x
Cell internal structure	x	x	x	x			x				
WARNING of persons AND correct reaction of persons					x	x				x	x

Fig. 1.3 Example for malfunction *Overcharge* Mapping of safety measure to battery level and safety disciplines

context in which the battery will be used. These assumptions have to be documented and considered for use. Appropriate safety measures have to be applied in the vehicle, in order to prevent malfunctions in the battery.

Sample safety measures:

- [ORGA] + [MECH] + [CHEM] + [ELEC] Fire extinguisher inlet—The fire extinguisher inlet of the battery system has to be reachable for the fire brigade.
- [MECH] + [THERM] Thermal management—Cooling and heating of the cells, as requested by the battery system.
- [ORGA] + [MECH] + [CHEM] Vent gas management—The vehicle must contain adequate outlet for the vent gas in case of a cell defect.
- [FUSA] Operational Strategy—The vehicle should manage the driving strategy of the powertrain, and critical situations should be prevented by an overall vehicle safety concept (e.g. overcharge, over-temperature).
- [FUSA] Warning concept—People in and around the car should be warned by visual and acoustic signals.

Figure 1.3 shows an example for the malfunction *Overcharge*. It provides an overview of possible safety measures and their mapping on the entities of the battery system and on the different safety disciplines.

1.4 Considering non-E/E Measures in the Concept Phase

So far, we have seen that functional safety is just one aspect that has to be considered for the development of a safe automotive system. In this section, we describe a modified version of the ISO 26262 safety workflow, which consists of 3 main activities. Below, these activities and our proposed modifications are described in more detail

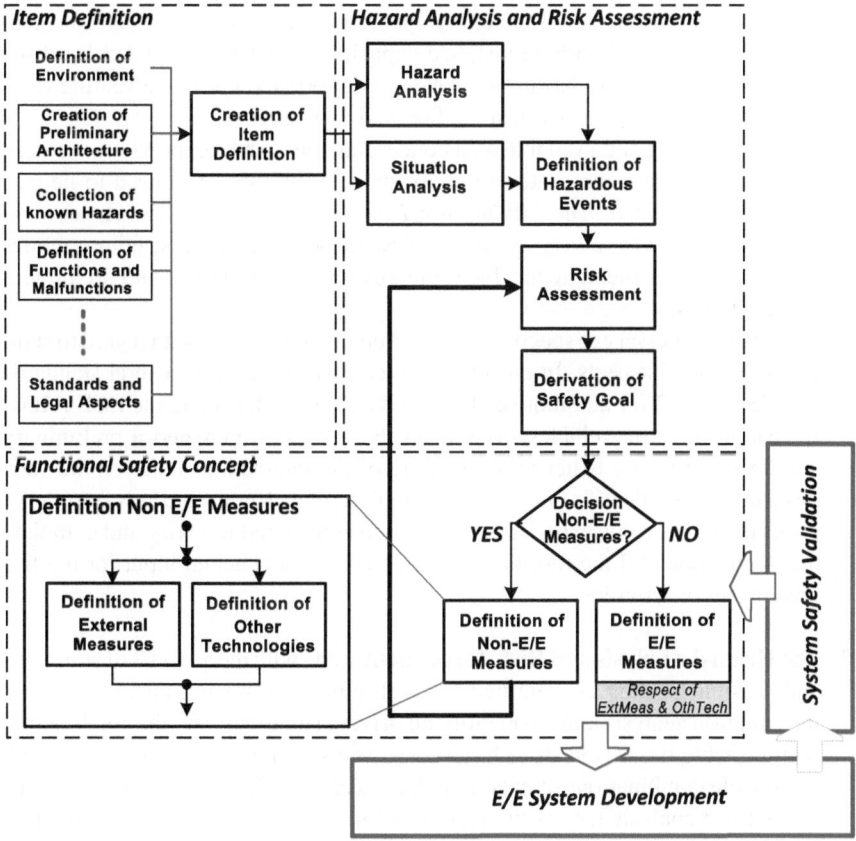

Fig. 1.4 Workflow of the concept phase following ISO 26262, including item definition, hazard analysis and risk assessment, and the functional safety concept. *Note* A proof of the controllability parameter, which is needed during the safety validation, is not illustrated in this figure

using the example of an HV lithium-ion battery. This work was conducted in an internal project, and the workflow has previously been published in SEAA2013 [7].

The main purpose of the modified workflow is the holistic investigation of safety measures from different disciplines at an early stage of development. This means that non-E/E measures are already considered in the concept phase, whereas the original workflow sees them as an add-on in later phases.

Basically, three main activities are considered here, as illustrated in Fig. 1.4: (1) Item definition, (2) Hazard Analysis and Risk Assessment, and (3) the design of the Functional Safety Concept. Below, these activities and the newly introduced iteration loop are described.

1. **Item (system) definition**, the first activity in the concept phase, starts with the definition of the item—the system, its functions on vehicle level, and its boundaries to other items. The item in this example is an hazardous voltage

(HV) lithium-ion battery. The battery should be used in a Plug-in Hybrid Electric Vehicle (PHEV) with an installed capacity of 24 Ah. Potential risks of the lithium-ion battery are hazardous voltage (U > 60VDC), leakage/venting (corrosive/toxic, flammable, explosive), fire, and explosion.

First, all relevant and available data concerning the item (e.g. previous projects, customer requirements, state-of-the-art, market analysis, etc.) need to be collected and analyzed. The *Lithium-Ion Batteries Hazard and Use Assessment Report* [9] provides a very good overview of possible hazards, failure modes and hazard assessment, applicable standards for the US market, and fire protection strategies.

It is further necessary to specify non-functional requirements with regard to standards and legal aspects. In our basic project, we scrutinized several standards (e.g. ISO 26262 for automotive electric/electronic systems and the ECE R100[7] for battery electric vehicles). Based on the results, we created a preliminary architecture to get a better understanding of the interactions between the various parts and to identify functions and malfunctions. Known hazards from other projects and previous experiences have been considered to verify and complete the description. All the results of this step are a fundamental input for the following safety activities.

2. The **Hazard Analysis and Risk Assessment** starts with the analysis of situations and possible hazards, as identified in a preliminary hazard analysis. The following situation analysis aims to identify all driving situations, and the combination with possible hazards leads to hazardous events. Driving situations contain all reasonable combinations of operational, environmental, and weather conditions. The hazard analysis targets the identification of potential hazards for the item on the top level of the system.

We used a Con-FMEA, a systematic method recommended by ISO 26262, to identify the potential hazards of the HV battery system. This approach provides support for traceability, the possibility to verify the completeness of the hazard analysis, and the extension of the Con-FMEA for other FMEAs in the following development phases, as shown in Fig. 1.5. This means that the causes of the failure modes of the Con-FMEA form the new failure modes for the System FMEA. The connections between the identified hazards and the different kinds of failures at different levels of development builds up a complete failure net. This failure net is a step-by-step refinement in the FMEA, which supports failure propagation and traceability.

In our example, the hazard and situation analysis resulted in 640 hazardous events. These hazardous events were identified by a stepwise combination and filtering of possible combinations of operational, environmental, and weather conditions. Finally, the plausibility of each combination was checked. As a result, we identified 121 plausible hazardous events, which were then assessed

---

[7] ECE R100—Uniform provisions concerning the approval of vehicles with regard to specific requirements for the electric power train.

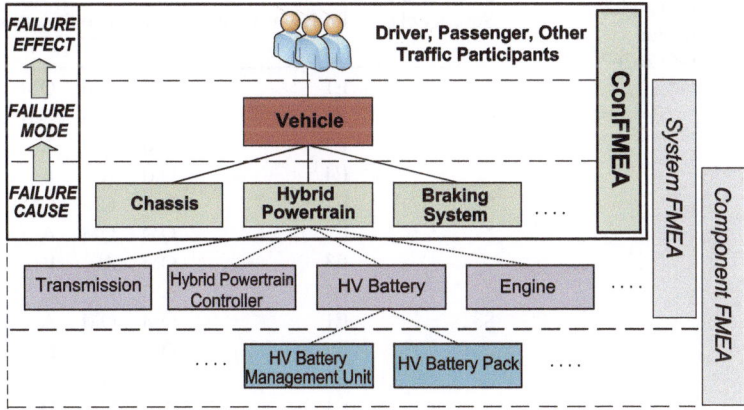

Fig. 1.5 FMEAs applied on different development levels

according to the risk assessment parameters *Severity (S)* [S0..S4], *Exposure (E)* [E0..E4], and *Controllability (C)* [C0..C3]. If any of these parameters results in a "S = 0 OR E = 0 OR C = 0" no safety development is needed—the level QM (quality management) is sufficient. The rationale behind each classification has to be documented appropriately because it is the basis for the ASIL determination, according to the risk graph of ISO 26262 (see Table 1.1).

Finally, safety goals have to be specified depending on the hazardous events and risk assessment results.

Below, an exemplary classification of a hazardous event is shown, where the vehicle is deactivated:

- **Hazardous event:** *Fire because of internal cell defect during parking situation (system is deactivated)*
- **Individuals at risk:** *Persons around the vehicle (Assumption: There is noone in the vehicle during the parking situation.)*
- **Possible harm:** *Burning of cell could cause hot smoke gas that could lead to smoke gas contamination and burns of critical injury degree are possible.*
- **Perception:** *Unpleasant sweet smell, and visible smoke*
- **Severity:** *S2— Severe injuries possible (life-threatening, survival probable)*
- **Exposure:** *E4—The vehicle will park every day for a long time in the parking garage.*
- **Controllability:** *C3—Less than 90 % of all drivers or other traffic participants are usually able, or barely able, to avoid harm.*

One main challenge here is the fact that the E/E system of the PHEV is deactivated during parking. For this specific situation, it is not possible to fulfill the safety goals with E/E measures only because these measures mainly mitigate hazardous situations during operational modes.

Table 1.1 Risk graph for ASIL determination according to ISO 26262 [Part3]

Severity class	Probability class	Controllability class		
		C1	C2	C3
S1	E1	QM	QM	QM
	E2	QM	QM	QM
	E3	QM	QM	A
	E4	QM	A	B
S2	E1	QM	QM	QM
	E2	QM	QM	A
	E3	QM	A	B
	E4	A	B	C
S3	E1	QM	QM	A
	E2	QM	A	B
	E3	A	B	C
	E4	B	C	D

We derived the required ASIL for our exemplary hazardous event using the risk graph (Table 1.1) of ISO 26262 : *Severity S2*, *Exposure E4* and *Controllability of C3 → ASIL C.*

The last step is the derivation of safety goals, as in this case *"Avoidance or/and mitigation of hazards caused by internal cell defect."* with the safe state *"No fire outside of the vehicle."*

3. The **Functional Safety Concept (FSC)** describes the derived safety measures (see Fig. 1.6) which realize the safety goals. Following ISO 26262, there are three different types of safety measures (E/E safety measures, other technologies and external measures). One viable approach to fulfill the safety goals in this case is the consideration of non-E/E measures in order to reduce the required ASIL.

Our modified workflow introduces an additional decision regarding whether or not it is possible or better to define non-E/E measures to fulfill the safety goals. If it is, we propose the identification of non-E/E safety measures with support from specialists from other disciplines (e.g. mechanical engineering). They need to be involved at an early stage of development because, based on their expertise, external measures and other technologies can be elaborated and considered. An example of another technology measure for an HV battery is the use of *fire-resistant materials for the battery housing* and an external measure could be a *fire detector in the parking garage*. All kinds of safety measures have to be introduced in the FSC as Functional Safety Requirements, which are linked to the corresponding elements of the Functional Safety Architecture. The main elements of the identified E/E measure are a sensor, a processor and an actuator. The FSC should provide a safety event chain from the detection of critical signals (sensor) to the processing and correct decision for the safe operation (processor), and finally the execution of a safe state (actuator) defined in the top level safety goal for the specific hazardous event.

Fig. 1.6 Principle of functional safety concept consisting of three types of safety measures

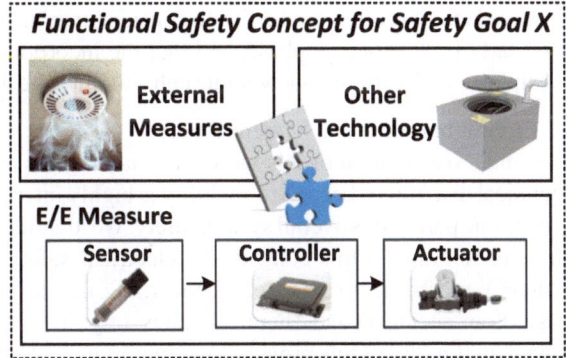

4. **Iterative refinement step, including update of functional safety concept:** After applying the different safety measures, we introduce a feedback step to repeat the risk assessment with the new conditions. The following measures were defined for the HV battery example: (1) External measures: *Fire detection unit and fire extinguisher have to be installed in the parking garage*, and (2) other technologies: *Fire-resistant housing of the battery system.*
   The introduction of these measures changes the risk assessment as follows:

   - **Severity:** *S1—Light or moderate injuries possible (not life-threatening)*
   - **Rationale for new S:** *The use of special fire-resistant materials for the mechanical construction of the housing will reduce the intensity of the harm.*
   - **Controllability:** *C2—90 % or more of all drivers or other traffic participants are usually able to avoid harm.*
   - **Rationale for new C:** *People will be warned by acoustic signals from the fire detection unit; a fire extinguisher will be available to extinguish the fire; the fire brigade will be alerted in the case of fire.*

   This leads to the new rating result of an ASIL A classification for the hazardous event. Lowering the required ASIL from *C* to *A* means that the remaining risk which has to be covered by E/E safety measures is lower, and therefore a less complex E/E measures and less development effort are needed.
   The last step is an update of the functional safety concept. Each introduced safety measure that contributes to the risk reduction is specified as functional safety requirements, which are mapped to the elements of the functional safety architecture. See Fig. 1.6 for the main parts of the functional safety concept.

1.5 Discussion and Conclusion

This chapter summarized our investigations of functional safety based on ISO 26262 for HV batteries typically used in EVs or HEVs. We presented an approach for an iterative determination of the required ASIL by applying non-E/E measures. We observed that it is often productive to consider external measures and other technologies early in the concept phase, and that the incorporation of different engineering disciplines with different viewpoints helps to improve the safety of the entire system.

Functional Safety $\subseteq$ System Safety

One main observation of this work was that hazards and risks result from different technical disciplines because e-mobility is highly interdisciplinary. Functional safety covers one part of this overall system safety. We identified several other types of safety that are relevant in this context, i.e. electrical safety (e.g. considering hazardous voltage), mechanical safety (e.g. concerning the deformation of the battery in the case of an accident) and chemical safety (e.g. helps to prevent explosion or fire). One main finding of this project is the importance of a strong interaction of all these different safety disciplines in the concept phase, which requires an organizational safety culture that fosters interaction between different disciplines. Not all hazardous events can be covered by E/E safety measures alone. Other technologies or external measures are equally important in order to achieve a safe system state.

Intercultural Aspects

The discussion with other departments results in a more holistic, interdisciplinary system and safety understanding. It also reveals how each team is able to contribute to the safety of a system. A discussion at an early stage of the project definitely improves the interaction between the different teams. Nevertheless, it has to be kept in mind that different views include different opinions, and often even contradicting opinions. All of them are correct in their specific systems or safety views. This can result in never-ending discussions, if there is no clear moderation.

We can offer one example of a discussion about the definition of the safe state of the system. One common function of the battery is *charging*. In the case of *overcharge*, the engineers responsible for electrical safety define the protective safe state in any such case for the electrical system to disconnect the HV battery from the HV net of the vehicle. This would lead to an undefined operation condition of the vehicle. The functional safety team must think about any possible driving situations, where an unintended loss of HV energy could lead to a critical situation. One such situation could be an overtaking maneuver on a country road, where a significant loss of driving torque could lead to a dangerous situation for the driver or other traffic participants.

Scope of Functional Safety

With a holistic safety view, it is often difficult to define the responsibilities for different hazards. Sometimes hazards are not directly caused by an E/E failure, but are an indirect consequence of a malfunctioning E/E system. Regarding the example of electric or hybrid electric vehicles, it cannot be clearly defined whether or not the HV battery should be considered only as an E/E system.

Acknowledgments The authors would like to acknowledge the financial support of the "COMET K2—Competence Centres for Excellent Technologies Programme" of the Austrian Federal Ministry for Transport, Innovation and Technology (BMVIT), the Austrian Federal Ministry of Economy, Family and Youth (BMWFJ), the Austrian Research Promotion Agency (FFG), the Province of Styria and the Styrian Business Promotion Agency (SFG).

The research leading to these results has received funding from the ARTEMIS Joint Undertaking under grant agreements pSafeCer no 269265 and nSafeCer no 295373. Further, the authors would like to acknowledge APIS GmbH for their support by their software tool, APIS IQ FMEA PRO.

References

1. Clifton AE et al (2005) Hazard analysis techniques for system safety. Wiley.com, New York
2. Ford Motor Company (2004) FMEA Handbook Version 4.1
3. IEC 61508 (2010) Functional safety of electrical/electronic/programmable electronic safety-related systems, 2nd edn. International Electrotechnical Commission, Geneva
4. ISO 26262 (2011) Road vehicles - Functional safety International Standard, parts 1–10. ISO copyright office
5. Leveson N (ed) (1995) Safeware system safety and computers. Addison-Wesley Publishing Company Inc, New York
6. Mader R et al (2011) A Computer-Aided approach to preliminary hazard analysis for automotive embedded systems. In: 18th IEEE international conference and workshops on engineering of computer based systems (ECBS)
7. Martin H et al (2013) Investigation of the influence of non-E/E safety measures for the ASIL determination. In: 39th EUROMICRO conference on software engineering and advanced applications (SEAA)
8. Mehrdad E et al (2011) Modern Electric, Hybrid Electric, and Fuel Cell Vehicles: Fundamentals, Theory, and Design. CRC Press, Boca Raton
9. Mikolajczak C et al (2011) Lithium-Ion Batteries Hazard and Use Assessment. Technical representative, Exponent Failure Analysis Associates, Inc./ Fire Protection Research Foundation, Final Report
10. UN Recommendation (2009) UN Recommendations on the Transport of Dangerous Goods, Manual of Tests and Criteria 38.3 Lithium batteries, Rev. 5, Amend.1

Chapter 2
Battery Modelling for Crash Safety Simulation

Gernot Trattnig and Werner Leitgeb

Abstract Finite element battery models used for crash simulation are effective tools for designing safe, lightweight battery systems for electric and hybrid electric vehicles. This chapter describes the currently available methods for integrating batteries into full-vehicle crash models and discusses their limitations at the present state of implementation. Innovative modelling approaches are able to determine the specific battery failure modes, such as short circuits and (electrolyte-) leakage. These methods are discussed and evaluated here based on their future applicability in the vehicle design process.

Keywords Finite element method · Crash simulation · Battery crash safety · Battery deformation and failure · Jelly roll

2.1 Introduction

Due to the conventional areas of application for lithium-ion batteries (e.g. mobile phones or laptops), battery research and the corresponding development of novel modelling techniques has focussed primarily on goals such as improved capacity, power and durability. This is also the main expertise of the battery producers and the associated scientific community. With the increased application of lithium-ion batteries in modern electric vehicles (EV) and hybrid electric vehicles (HEV), the requirement of crash safety has become important. Therefore, the automotive industry requires highly predictable, applicable and efficient methods for simulating battery deformation and failure in crash test situations.

G. Trattnig(✉) · W. Leitgeb
Virtual Vehicle Research Center, Graz, Austria
e-mail: gernot.trattnig@v2c2.at

W. Leitgeb
e-mail: werner.leitgeb@v2c2.at

A. Thaler and D. Watzenig (eds.), *Automotive Battery Technology*,
Automotive Engineering: Simulation and Validation Methods,
DOI: 10.1007/978-3-319-02523-0_2, © The Author(s) 2014

2.1.1 Motivation

The demand of electric energy for high vehicle ranges in HEVs and EVs results in batteries with weights of up to several hundred kilograms and considerable volumes. Since the deformation of the battery can lead to hazardous situations, one aim of the current vehicle development is to prohibit any significant deformation of the battery in crash tests. This can only be achieved by tightly restricting the available space for the battery system and high—but heavy—stiffness of the battery pack.

In order to enable the development of long-range, lightweight EVs, the engineer needs a better understanding of the battery deformation and failure characteristics, as well as new simulation tools. These tools must have the same accuracy and reliability as the numerical vehicle development methods in use today. In this way, it will become possible to develop structural battery concepts with optimal use of the available space at minimum weight and with increased crash safety.

2.1.2 Specific Hazards of Electric Vehicles

Crash safety for batteries means that an accident does not cause dangerous voltages, vent gas, heat or fires, which could harm the environment, passengers, pedestrians or rescue teams. This can be accomplished by the battery design itself, together with structural protection measures implemented during the vehicle integration.

Hazardous voltages of 400–800 V can lead not only to human injury, but also to short circuits and arcing, which can generate heat and trigger additional failure modes in the battery system.

Short circuits within the battery cells' active material or due to contact of conducting components with different potentials can cause electrolyte gas to develop and can lead to degassing or the leakage of cell-internal fluids. These vent gases and liquids are flammable and possibly toxic and therefore must not come into contact with passengers.

The worst-case scenario in the car crash is the combination of vent gas or leaking fluids and ignition points, such as arcing or hot spots. This combination can lead to fires and exothermal reactions in the cell itself, with unpredictable consequences for trapped passengers. As an illustrative example, Fig. 2.1 shows the exothermal reaction of a single charged lithium-ion metal-oxide cell caused by severe deformation under laboratory conditions.

2.1.3 Applicable Design Approach for Batteries

In order to design crash-safe batteries for EVs and HEVs, validated and highly predictive battery models are needed in the development process. They must describe

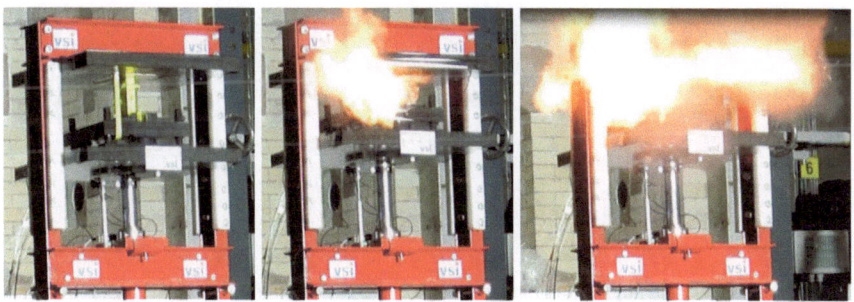

Fig. 2.1 Exothermal reaction of a single charged cell under severe deformation—test conducted in cooperation with TU Graz, Vehicle Safety Institute

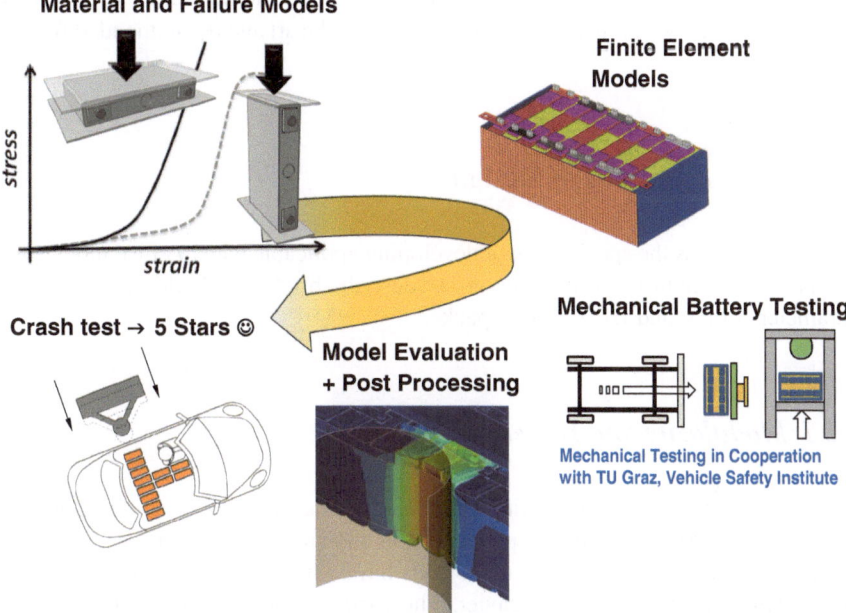

Fig. 2.2 Suggested development approach for validated finite element battery models used for the design of crash-safe electric vehicles

deformation, mechanical and electro-chemical failure and have to be applicable in the current car crash finite element (FE) models.

Figure 2.2 shows the steps suggested for the development of a validated FE model of a battery. The first step is the mechanical testing of a battery cell. This enables the build-up of suitable models for the single cell, with characteristic deformation and failure behaviour. Battery module or pack models can then be created by applying state-of-the-art FE techniques. The derived models must be validated in specially designed battery module or pack tests.

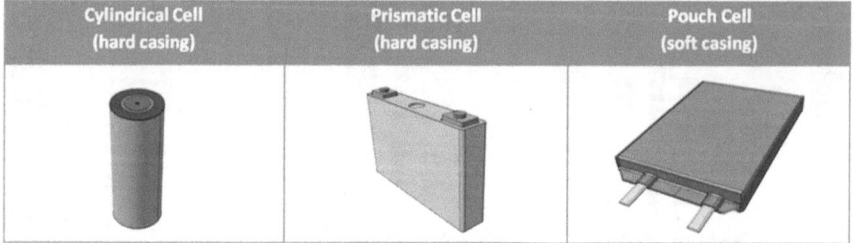

Cylindrical Cell (hard casing)	Prismatic Cell (hard casing)	Pouch Cell (soft casing)

Fig. 2.3 Schematic drawings of the main cell types used in the automotive industry

This chapter describes the boundary conditions of the vehicle development process, the required tests and the individual steps for the derivation of a battery model, followed by summary of the current state of the art and recommended further development.

2.2 Automotive Battery Design

In order to discuss the special task of developing applicable battery crash models for the automotive industry, it is necessary to describe briefly the build-up and design parameters of EV and HEV battery packs.

2.2.1 Modularity and Battery Components

Battery cells are the smallest unit in the battery. The three common types are the cylindrical, the prismatic and the pouch cell, as shown in Fig. 2.3. Due to their sheet metal casing, cylindrical and prismatic cells have a higher structural integrity than pouch cells, but they are also heavier. The casing is often made of quite strong aluminium sheets, in contrast to the polymer, coffee bag like, cover of the pouch cell.

The main component of the cell is the active material, often referred to as jelly roll.

Other components of a working battery cell are current collectors and terminals, the aforementioned cell casing, spacers and isolators within the casing, and a safety pressure valve. A lithium-ion cell usually features a voltage of about 2.5–4.2 V between the two terminals, depending on the chemistry, the load situation and the state of charge (SoC). Since powerful electric vehicle motors work at voltages of about 200–800 V to be efficient, several hundred cells in series connection are needed to provide this elevated voltage. Cells are grouped to modules for several mainly practical reasons, including relatively low voltages ($<$ 60 volts), sizes and weights that can be handled by a single worker, and modularity.

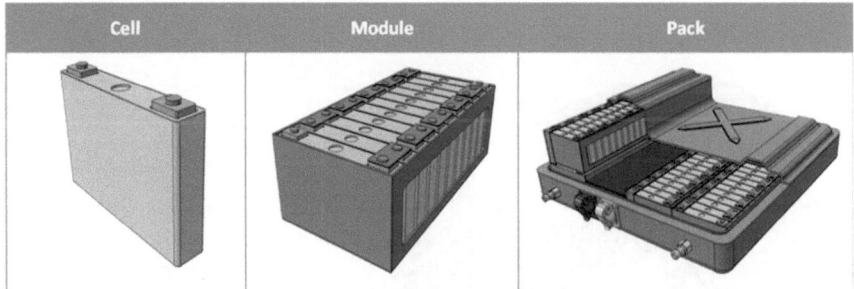

Fig. 2.4 Drawings of the modular parts of a battery: a single battery cell (*left*), a battery module (*middle*) and the complete battery pack (*right*)

Modularity helps reduce the amount of different parts within a battery pack and allows packs to be designed using the same basic modules to handle different energy content, voltage and designs requirements. The battery pack contains all the cells and modules of the battery. It also usually contains the cooling part of an environmental system to keep the cells within their admissible temperature limits, a battery management system (BMS), and its associated hazardous voltage (HV) protection system. Figure 2.4 shows drawings of the modular parts of a battery.

Since safety is a mandatory requirement, the pack is hermetically closed, with degassing vents leading possibly dangerous electrolyte gases away from the passenger compartment. The battery system, as the highest integration level, contains the battery pack and all electrical cables, sockets, and sensors distributed throughout the vehicle, which are needed to run the battery in the vehicle environment. Thus, within the battery system, component size spans several orders of magnitude, from single layer active cell materials of 1/100th of a millimetre thickness to the battery pack of 1–2 m in width and length and several hundred kilograms of weight.

2.2.2 Safety-Relevant Design Parameters

In order to enhance crash safety, dangerous conditions causing short circuits or contact of cathode and anode due to separator damage must be avoided, as they can lead to hot spots and subsequent electrolyte decomposition, heat and vent gas generation. Therefore, one must examine the main influencing factors that determine the hazards in the battery and facilitate their consideration during the design process. The design parameters at the cell level are very much constrained by electro-chemical design requirements, and the chemical reactivity of the jelly roll or active material depends strongly on the chemistry used. Soft or hard casing and the form factor [1] have a strong influence on the module design and module failure characteristics. For the battery module or battery pack, the introduction of crash safety features is a focus of the development. Both crash and transport safety can be improved by the

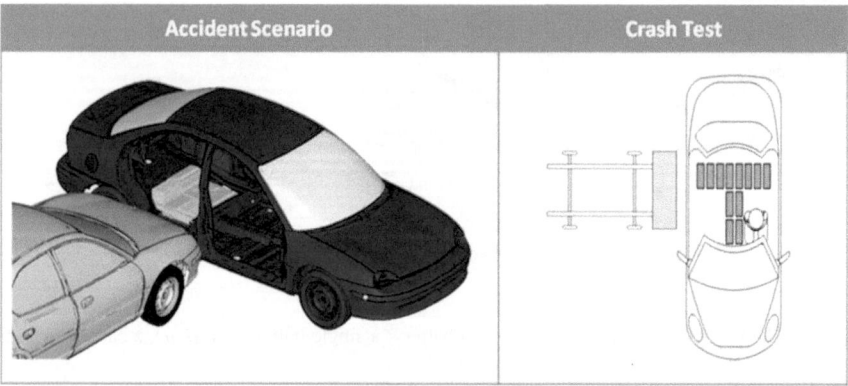

Fig. 2.5 Non-standardized crash test with an EV; FE model of the crash test with masked components for visibility of the battery under the back seat (*left*), schematic drawing of the crash with possible battery positions (*right*); FE Model courtesy of National Crash Analysis Center (NCAC)

appropriate design of casing, joints and isolators. The aims are to avoid possible contacts of electric conductors and to restrict the deformation of the battery cell to uncritical levels.

Finally, the choice of the battery pack geometry and position and the structural design of the vehicle are the main safety-relevant design parameters when integrating the battery pack in the vehicle.

For the design, it is a safe way to prohibit any deformation of the battery itself in order to eliminate the possibility of any hazardous event. Therefore, the batteries are grouped in structurally stiffened and reinforced compartments in the vehicle, where no deformation is expected in standardized crash tests (Fig. 2.5).

2.3 Structural Vehicle Design Process Including Batteries

This chapter gives a short overview of the modern structural vehicle design process and its dependence on FE simulation. The proposed methods are described, and the performance specification for a FE battery model is defined.

Modern vehicles are designed according to many different requirements. Apart from the obvious ones (e.g. saleability, through exterior and interior design or performance and drivability), one very important and legally binding aspect is the vehicle's safety performance in an accident, as schematically shown in Fig. 2.5. The focus is on protecting the individuals involved and reducing accident-related injury. The laws differ from country to country, but generally the United States (US) FMVSS[1] and the European ECE[2] regulations form the basis. On top of these laws, widely accepted

---

[1] FMVSS: Federal Motor Vehicle Safety Standards.

[2] ECE: Economic Commission for Europe.

consumer test procedures enhance the safety requirements even further. In Europe, this is the Euro NCAP consortium[3] and several smaller national organisations, as well as companies such as the German ADAC[4] or the British Thatcham Research. Modified NCAP programs are also used in China, Australia, Brazil, the US and Japan. In the US, the IIHS[5] establishes additional performance criteria. Common to all these tests is that standardized full-vehicle crash tests that simulate the most common and dangerous real-world accidents must be performed under strict predefined conditions in order to rate and compare the vehicles performance regarding vehicle safety. It is common practice for OEMs to strive for good results in these consumer tests, as they are widely known and respected.

2.3.1 Standard Approach and Requirements

In order to cope with this variety of requirements from legislative and consumer tests and to accelerate development time, simulation methods are used throughout the vehicle design and development process [2]. For structural integrity calculation and crash simulation, explicit FE methods [3] are normally used. Several crash solvers are commercially available. The most common ones are Abaqus, LS-Dyna, Pam-Crash and Radioss.[6]

Although usually cheaper than full-scale crash tests, crash simulations are limited by the costs of computer power. Since calculation time in explicit FE solvers depends on element number and size, only structurally important and necessary components are normally included in the model. As computer power increases, more detailed and better results can be obtained. The FE mesh of a full-vehicle model can therefore easily surpass 2 million calculation nodes and elements, with a characteristic length of between 2 and 10 mm, with 4–5 mm being the current standard. With the introduction of detailed battery models, node and element numbers will increase significantly.

2.3.2 Batteries in Crash Tests and Crash Simulation

As of 2013, a combination of transportation laws and recommendations[7] are used to rate battery safety in traction-battery-equipped vehicles, and standard crash tests must also be passed. However, battery cells show uncritical mechanical deformation potential in specially designed tests. To use this potential, it is necessary to fully understand the mechanical deformation and failure behaviour of batteries. FE battery

---

[3] NCAP: New Car Assessment Program.

[4] ADAC: Allgemeiner Deutscher Automobil-Club e. V.

[5] IIHS: Injury Institute for Highway Safety.

[6] SIMULIA Abaqus FEA, LSTC LS-Dyna, ESI Group PAM-Crash, Altair Engineering RADIOSS.

[7] 38.3 Drop Tests [4], FreedomCAR [5], EUCAR hazard levels.

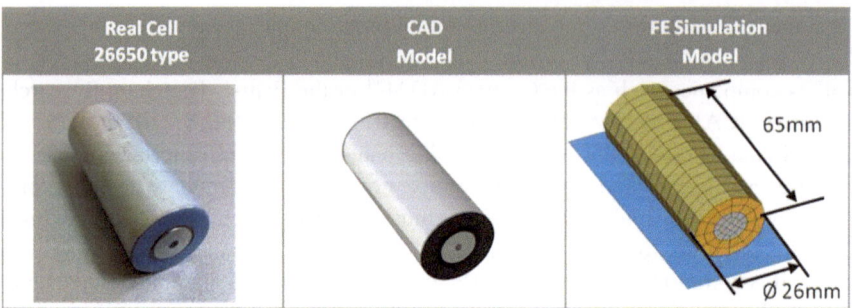

Fig. 2.6 Picture of a cylindrical cell with an aluminium casing (*left*), the CAD model of the cell (*middle*), and the FE model (*right*)

models, which must be able to depict this behaviour, are becoming essential for optimising location and structural reinforcement for an acceptable cell deformation.

2.4 Finite Elements Model of the Battery

The integration of the battery pack in crash-safe electric vehicle development also means integrating the battery model into the crash simulation, including all components that are structurally relevant for the battery. This can be done best by using the already established explicit finite element solvers and methods and adapting them where necessary.

FE solvers for full-scale vehicle crashworthiness simulation are limited by element size and time step in order to maintain a manageable model size and thereby keep the calculation time within manageable limits. Generally, FE models are derived from complete three-dimensional computer aided design (CAD) models that accurately represent the real object. Construction drawings can be derived directly from these CAD models. Generally, an FE geometric model mesh is composed of one-dimensional bars and links, two dimensional sheet-like structures and three-dimensional volume components [3]. The reduction of geometric details is one of the constraints when building an FE model, as details smaller than 4–5 mm are omitted or replaced. As an example, Fig. 2.6 shows the differences between a cylindrical cell and its CAD and FE models, and Fig. 2.7 shows the individual components of this cell and the corresponding parts in the FE simulation.[8]

The mechanical description of all structurally important battery components is done in the same way as for conventional ones, that is by using a node and element-based geometry, superimposed with stress-strain curve-based material models. For the simulation of other current carrying components (e.g. busbars and HV cables),

---

[8] All images of cylindrical cells in this chapter show type 26650 cells (26 mm diameter and 65 mm length).

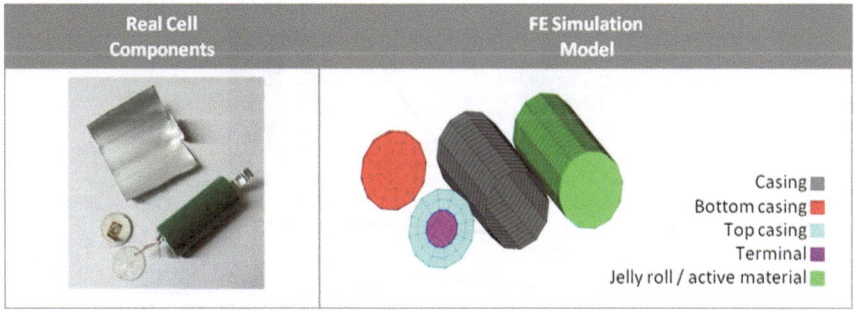

Fig. 2.7 Picture of the battery cell components of a cylindrical cell with an aluminium casing (*left*) and the corresponding parts of the FE model (*right*)

new methods are needed for the modelling of deformation and failure. For all components, suitable material models need to be developed to adequately describe the mechanical behaviour of the different battery components. The following chapter describes and discusses the applicable methods.

2.4.1 Modelling of Mechanical Deformation

The basis of an accurate failure evaluation is the modelling of the deformation which causes the failure. This chapter briefly discusses available methods for the different battery components.

Battery pack: The main load-bearing component of the pack is the casing, which should be leak-tight. The casing can be made of sheet steel or lightweight materials such as aluminium or fibre-reinforced plastics. These materials are also found in the body-in-white structure, and various plasticity-strain-rate-dependent material models are available in the crash solvers [6–12]. The elastic deformation of connectors (e.g. spot welds, rivets or screws) can be modelled by link elements with corresponding elasticity parameters [7, 11].

Battery module: As in the battery pack, the deformation of the casing, conductors, isolators and joints can be modelled with standard FE methods. The main difference is a possible pre-loading of the modules, which is done in order to apply a constant pressure on the battery cells. This is necessary in order to ensure a high electrochemical lifetime of the active cell material. The pre-loading can influence the module's stiffness significantly. In this case, it is necessary to model the pre-loading process and map the elastic pre-deformation and pre-stresses on the crash model. This can be done by the available Forming to Crash methods in most common crash solvers [13].

Battery cell: The cell has very strong anisotropic deformation behaviour, as shown in Fig. 2.8 for a cylindrical cell. Depending on the cell type (Fig. 2.3), the casing can be important for the battery cell stiffness. Here again, available standard FE methods

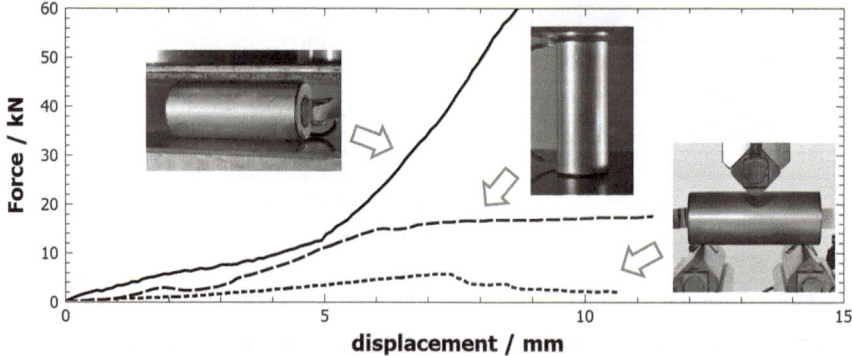

Fig. 2.8 Figure with anisotropic deformation behaviour of cylindrical cells; compression tests normal to cell axis (*solid line*) and in cell axis (*dashed line*) and 3-Point-Bending tests (*dotted line*)—test conducted in cooperation with TU Graz, Vehicle Safety Institute

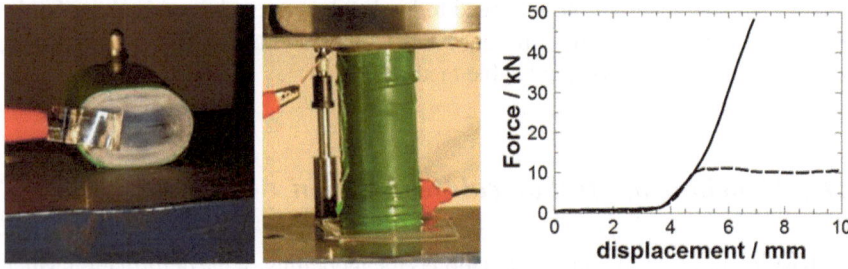

Fig. 2.9 Anisotropic jelly roll deformation of a cylindrical cell without casing; compression normal to the cell axis (*left*) and parallel to the cell axis (*middle*); (*right*) force versus displacement curves of normal (*solid line*) and parallel (*dashed line*) compression tests

are used to model the cell casing. At this level, relatively small features of the cell can also be important for their deformation and subsequent failure behaviour (e.g. current collectors in the cell and details of a cylindrical cell are shown in Fig. 2.13).

Here, it can be necessary to simplify the actual geometry, since an applicable FE crash net has a mesh size of about 5 mm, as shown for a cylindrical cell in Fig. 2.7. This can be done if the local deformation effects are understood and taken into account in the subsequent failure assessment.

The active material, the *jelly roll*, contributes to the cell stiffness. Depending on the loading direction, it can be a major load-carrying component with a strong anisotropic deformation behaviour (shown in Fig. 2.9).

In contrast to the casing materials and joints, the jelly roll itself is a new material in the crash simulation. Depending on the loading direction, mainly the porous active material (e.g. graphite, metal oxide or separator) or the conducting electrodes (e.g. aluminium or copper foils) are compressed and contribute to the cell stiffness.

There are two different approaches to this problem. The bottom-up approach is based on the idea of modelling the individual layers with their appropriate material

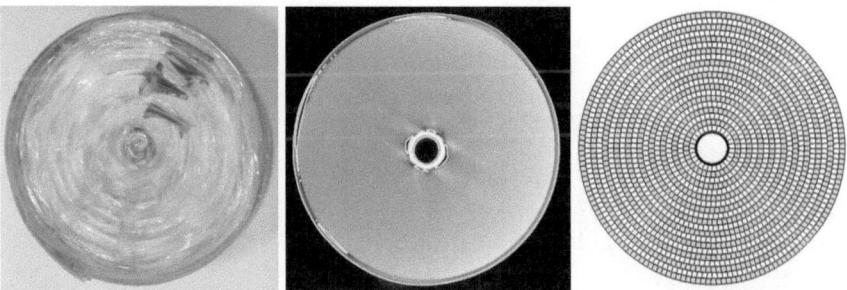

Fig. 2.10 Photo of a cylindrical jelly roll with a steel tube in the *centre* and a diameter of 26 mm (*left*), an X-ray tomography image of the cross section (*middle*) and the associated detailed finite element model (*right*)—X-ray tomography by the Austrian Foundry Research Institute (ÖGI), Leoben, Austria

behaviour [14]. Figure 2.10 shows cuts through a cylindrical jelly roll and a detailed model as an example, although not every single layer is modelled, the discretisation allows the investigation of the microscopic deformation behaviour. The fine mesh, necessary for this method, leads to high calculation times that are not acceptable for the crash simulation. Another problem is the measurement of the material data of the thin metal sheets, the electrolytes, the separator and the porous active material. Since the measurement is quite complicated, the mechanical properties are partly unknown or only available for different testing conditions (e.g. higher sheet thicknesses or different electrolyte levels). This approach is a more scientific one, which is suitable for investigating the deformation mechanisms in the cell and for deriving the macroscopic deformation behaviour from the jelly roll structure.

One applicable top-down approach is based on a macroscopic model of the jelly roll [15, 16]. Substitute models are used for the jelly roll in the crash model. For the parameterization of the model, the anisotropic deformation behaviour is measured by tests on the jelly roll or on individual battery cells (Fig. 2.11). Available honeycomb material models [7, 11] offer the ability to define the stress-versus-strain curves for each direction separately. The resulting model, which can describe the external deformation behaviour and deformation forces, is applicable in the crash simulation. Nevertheless, it does not describe the internal jelly roll deformation mechanisms and therefore cannot be used for the microscopic failure assessment.

2.4.2 Modelling of Material and Joint Failure

The failure assessment is based on an accurate description of the plastic deformation of the battery system's components and on the loads applied to the joints (Fig. 2.12). The mechanical failure has to be described, since it can lead to leakages (e.g. if the casing of a cell ruptures) or to a significant change in the deformation characteristic (e.g. if a load carrying component or a joint fails).

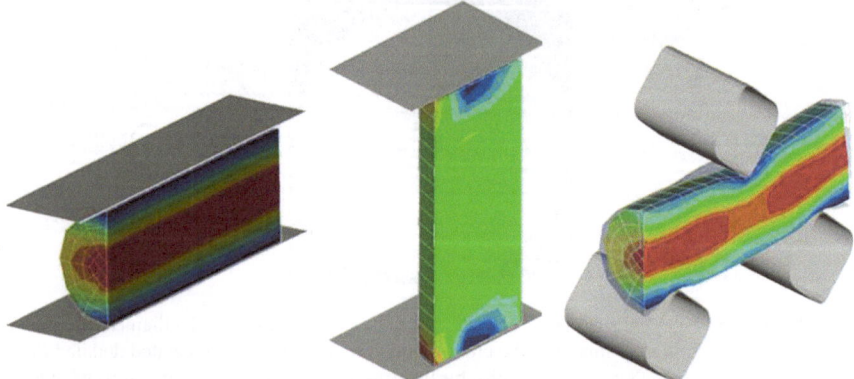

Fig. 2.11 Half finite element models of cylindrical cells with aluminium casing; compression tests normal to cell axis (*left*) and in cell axis (*middle*) and 3-Point-Bending tests (*right*)

Fig. 2.12 Deformation and failure of compressed cylindrical cells with aluminium casing; comparison of experiment and FE-model wrinkle formation in axial compression (*left*) and failure of a joint line in compression normal to the cell axis (*right*)

Various fracture models are available for describing the failure of *metal sheets*. Most of these calculate a damage value based on the plastic strain weighted by functions of the stress state [11, 17–20]. If the critical areas (e.g. a part of the battery pack or cell casing) are loaded in tension, they will give quite accurate results. One still unsolved problem in the applied simulation is the failure due to the fracture mechanic mode III [21], which means shearing by loading in sheet-normal direction. This failure mode can appear if a relatively sharp and stiff component, which can be a part of the battery pack or an intruding object, cuts into the sheet metal and causes localized failure without major deformation of the surrounding area. This is a challenging task in crash simulation, and novel element models with promising solutions are currently under development [11, 22].

For modelling *composites and isolators*, one must consider that, depending on the polymers used, they can be more brittle than the sheet metals in use. Due to the absence of significant plastic deformation, stress-based criteria are more suitable for describing that failure mode. New failure models for composites and polymers are available and are a focus of current development [23, 24]. Here, the application of Forming to Crash [13] methods is even more important than in sheet metals, since the local material properties caused by the production process depend significantly on parameters such as local fibre or polymer chain direction [25].

The other main factor for the strength of the battery system is the *failure of joints*. Depending on the joining concept, a battery system can contain adhesives, spot welds, laser welds, screws, or rivets, for example. In recent years, the failure of joints has been an important research topic in crash simulation. Therefore, various models for adhesives [26] and single-point connections such as spot welds and screws [27–30] are available and ready to use (see Fig. 2.12).

For the *failure of the jelly roll*, as with the non-active battery components, the failure assessment is based on an accurate description of the deformation. Due to the jelly roll deformation, internal short circuits—between the electrodes or from an electrode to the casing—can lead to heat generation and exothermal reactions. Concerning the deformation modelling, there are two possible approaches to follow.

The first approach is the bottom-up or scientific approach, where detailed FE models are used to describe failure mechanisms (e.g. the fracture of electrode layers, critical contacts or delamination—examples shown in Fig. 2.13) [14]. This microscopic approach can support the understanding of the jelly roll failure mechanisms and the development of suitable macroscopic jelly roll material models. The main problem remains the measurement of the microscopic material or contact zone parameters in tests, which can replicate the conditions in the cell itself. Because various parameters (e.g. fracture strains and stresses of the electrolyte-soaked active materials and conductor foils) have to be derived e.g. from literature or complex tests, the simulation results have to be interpreted with great care.

The top-down approach, which is applicable in the crash simulation, assesses failure by the observed macroscopic deformation of the jelly roll. This deformation and the related electromechanical failure can be tested and measured quite accurately, compared to the underlying microscopic mechanisms. Thus, based on a series of tests with deformations similar to the crash loading, a failure model for a cell can be parameterized. This failure model can be implemented in the jelly roll material model (e.g. based on FE element stresses and strains) or evaluated in the post-processing process, e.g. critical outer deformations (see Fig. 2.14). The disadvantage is that this failure model is not a general solution, but rather is only valid for the specific cell type and loading conditions tested.

2.4.3 Modelling of Electrical Contact and Leakage

The jelly roll modelling introduced the first failure models, which are not implemented in standard FE solvers yet. However, these are not the only failure mechanisms that are currently lacking appropriate modelling techniques. The three

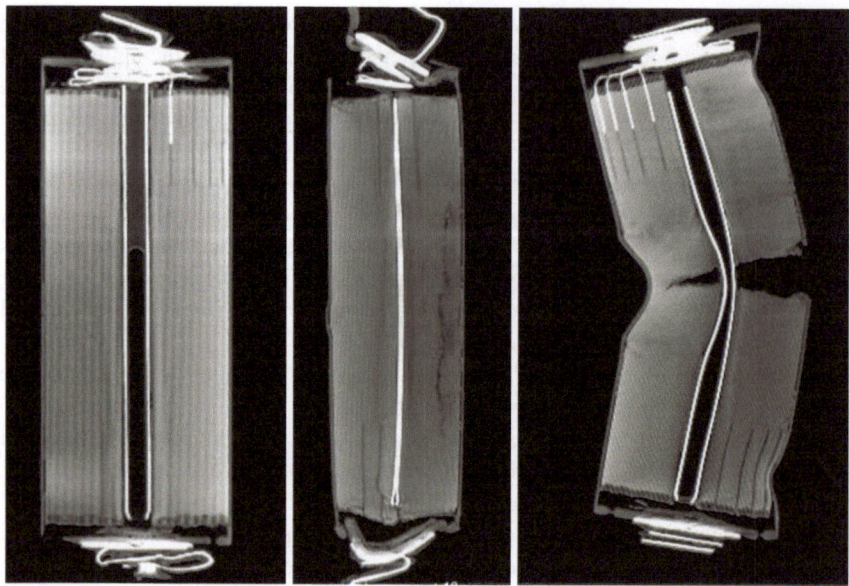

Fig. 2.13 X-ray tomography cross sectional images of cylindrical cell with a diameter of 26 mm of an un-deformed cell (*left*), a cell compressed normal to the cell axis (*middle*) and a cell deformed in a 3-Point Bending test (*right*)—X-ray tomography by the Austrian Foundry Research Institute (ÖGI), Leoben, Austria

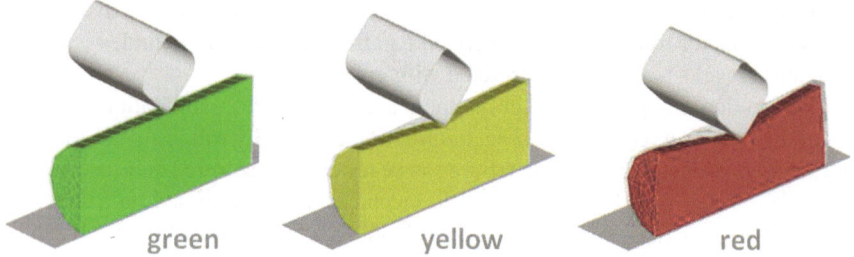

Fig. 2.14 Failure assessment of a cylindrical cell under compression based on the outer deformation; the colours indicate the criticality from *green* (uncritical) to *yellow* (critical) to *red* (failure)

additional main failure mechanisms are electric potential carryover, short circuits and leakage.

Hazardous voltages can emerge on bare conductive parts due to potential carryover, which is caused by contact with conductors following the crash deformation. Therefore, a risk analysis based on the components' potential difference and the contact situation is necessary.

In addition, short circuits due to failure of isolators and insulating layers are hazardous. For example, internal cell contacts from current conductors and casing,

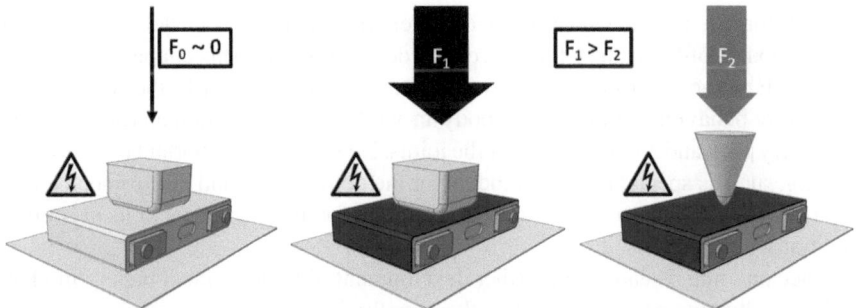

Fig. 2.15 Sketch of different possible short circuit situations between a battery cell and a conductive non-isolated metal impactor; short circuit between non-isolated can and blunted impactor without apparent force F_0 (*left*), isolated can and blunted impactor with high contact force F_1 (*middle*), and isolated can and sharp impactor with low contact force F_2 (*right*)

or an electrical contact between conductors and cell casing (see Fig. 2.15) can cause short circuits, which can lead to heat generation and exothermal reactions. To evaluate this risk, a detailed analysis of the contact situation in the FE simulation is mandatory, for example by evaluating the local pressures, taking into account the real local geometry (e.g. sharp edges) and the component's relative displacement. This difficult assessment of critical pressures and local geometries is not currently available in the crash solvers. Until detailed electrical contact models become available, a suitable post processing analysis is necessary.

Another hazard relevant for the post-crash safety analysis is the leakage of toxic electrolyte fluids and gas [31]. In order to ensure the sealing of the battery system, it is necessary to assess the integrity of the battery cell and pack casing. This can be done with methods for modelling the failure of the casings and joints such as laser welds, as discussed in Sect. 2.4.2, and an evaluation of the deformation and functionality of the seals and safety valves [32].

2.5 Conclusion

The crash safety requirements for lithium-ion batteries are currently met by avoiding any severe deformation on the battery pack, which is accomplished by limiting the available battery space in the car and by heavy structural protection measures in the vehicle.

This stands in strong contradiction to the design goal of increasing the range of electric vehicles by introducing high battery capacities and lightweight design. Thus, it is only possible to achieve this goal by allowing uncritical deformation to the battery (i.e. no heavy-weight battery packs and stiff components) and by developing new car concepts (i.e. optimal use of the available space).

Therefore, it is mandatory to develop reliable finite element deformation and failure models of the battery for the vehicle design process. This chapter has shown that the FE methods currently available and in use are able to describe the deformation and failure behaviour of the classic body-in-white structures and materials, such as the battery pack and module casing or the joints. Nevertheless, important tools are still missing, such as special material models for the deformation and electromechanical failure of the jelly roll, or electrical contact models for the assessment of local contact situations.

Since ongoing research and further development of finite element battery models is already showing promising results, these methods should soon become a standard tool in the vehicle development process.

Acknowledgments The authors would like to acknowledge the financial support of the "COMET K2—Competence Centres for Excellent Technologies Programme" of the Austrian Federal Ministry for Transport, Innovation and Technology (BMVIT), the Austrian Federal Ministry of Economy, Family and Youth (BMWFJ), the Austrian Research Promotion Agency (FFG), the Province of Styria and the Styrian Business Promotion Agency (SFG).

References

1. ISO/IEC PAS 16898:2012 (2012) Electrically propelled road vehicles—dimensions and designation of secondary lithium-ion cells
2. Kramer F, Franz U, Lorenz B, Remfrey J, Schöneburg R (2013) Integrale Sicherheit von Kraftfahrzeugen: Biomechanik - Simulation - Sicherheit im Entwicklungsprozess. ATZ/MTZ-Fachbuch
3. Bathe K (2002) Finite-elemente-methoden. Springer, Heidelberg
4. Recommendations on the transport of dangerous goods manual of tests and criteria (2009). Technical report, United Nations
5. Crafts CC, Doughty DH (2006) Sandia report FreedomCAR electrical energy storage system abuse test manual for electric and hybrid electric vehicle applications. Technical report, Sandia National Laboratories
6. Cowper G, Symonds P (1958) Strain hardening and strain rate effects in the impact loading of cantilever beams. Applied Mathematics Report, Brown University, Providence
7. ESI Group (2012) Virtual performance solution 2010
8. Hill R (1950) The mathematical theory of plasticity. University Press, Oxford
9. Johnson G, Cook W (1983) A constitutive model and data for metals subjected to large strains, high strain rates and hight temperatures. In: Proceedings of the 7th international symposium on ballistics, The Hague, The Netherlands
10. Jones R (1999) Mechanics of composite materials. Taylor and Francis, Washington
11. LSTC (2013) LS-Dyna manual
12. von Mises R (1913) Mechanik der festen Körper im plastisch-deformablen Zustand, Göttinger Nachrichten. Math Phys Klasse 4:582–592
13. Steinbeck-Behrens C, Steinbeck J, Schroeder M, Duan H, Hoffmann A, Brylla U, Kulp S, Pinner S, Rambke M, Leck L, Awiszus B, Bolick S, Katzenberger J, Schulz M, Runde S, Czaykowska A, Mager K (2012) Durchgängige Virtualisierung der Entwicklung und Produktion von Fahrzeugen (VIPROF). Technical report, BMBF, Germany
14. Sahraei E, Campbell J, Wierzbicki T (2012) Modeling and short circuit detection of 18659 Li-Ion cells under mechanical abuse conditions. J Power Sources 220:360–372

15. Greve L, Fehrenbach C (2012) Mechanical testing and macro-mechanical finite element simulation of the deformation, fracture, and short circuit initiation of cylindrical Lithium ion battery cells. J Power Sources 214:377–385
16. Wierzbicki T, Sahraei E (2013) Homogenized mechanical properties for the jellyroll of cylindrical Lithium-ion cells. J Power Sources 241:467–476
17. Bai Y, Teng X, Wierzbicki T (2009) On the application of stress triaxiality formula for plane strain fracture testing. J Eng Mater Technol Trans ASME 131(2):021 002–1–10
18. Basaran M, Wölkerling S, Feucht M, Neukamm F, Weichert D (2010) An extension of the GISSMO damage model based on lode angle dependence. In: LS-Dyna forum. Dynamore, Bamberg
19. Gurson A (1977) Continuum theory of ductile rupture by void nucleation and growth: Part I yield criteria and flow rules for porous ductile media. J Eng Mater-T ASME 99:2–15
20. Tvergaard V, Needleman A (1984) Analysis of the cup-cone fracture in a round tensile bar. Acta Metall 32:157–169
21. Anderson T (2005) Fracture mechanics—fundamentals and applications. CRC Press, Boca Raton
22. Kunter K, Heubrandtner T, Trattnig G, Mlekusch B, Fellner B, Pippan R (2011) Simulation of crack propagation in high strength automotive steel sheets using hybrid Trefftz method. In: 2nd European conference on eXtended finite element. Cardiff, UK
23. Knops A (2008) Analysis of failure in fiber polymer laminates: the theory of alfred puck. Springer, Berlin
24. Kolling S, Haufe A, Feucht M, Bois PD (2006) A constitutive formulation for polymers subjected to high strain rates. In: 9th international LS-Dyna users conference. Detroit, USA
25. Boisse P (2010) Simulations of composite reinforcement forming. In: Dobnik Dubrovski P (ed) Woven fabric engineering. InTech, Rijeka, p 387–414
26. P676: Methodenentwicklung zur Berechnung von höherfesten Stahlklebeverbindungen des Fahrzeugbaus unter Crashbelastung (2008). Technical report, Forschungsvereinigung Stahlanwendung e.V. Düsseldorf
27. Chauffray M, Delattre G, Guerin L, Pouvreau C (2013) Prediction of laser welding failure on seat mechanisms simulation. In: 9th European LS-DYNA conference. Manchester
28. Heubrandtner T, Scharrer G (2008) Hybrid-Trefftz formulation of spotwelds in car bodies. In: Leuven symposium on applied mechanics in engineering, pp 187–200
29. Malcolm S, Nutwell E (2007) Spotweld failure prediction using solid element assemblies. In: 6th European LS-Dyna users' conference. Gothenburg, Sweden
30. Szlosarek R, Karall T, Enzinger N, Hahne C, Meyer N (2013) Mechanische Prüfung von fliesslochformenden Schraubverbindungen zwischen faserverstärkten Kunststoffen und Metallen. Mater Test 10:737–742
31. Golubkov A (2013) Thermal-runaway experiments on consumer li-ion batteries with metal-oxide and olivin-type cathodes. In: RSC Advances
32. Brödner S (2012) Gummidichtungen in der Hydraulik - Grundlegendes und Möglichkeiten der FE-Simulation. In: 15. Poly-King Event, Würzburg

Chapter 3
Thermal Runaway: Causes and Consequences on Cell Level

Andrey W. Golubkov and David Fuchs

Abstract Lithium-ion batteries play an ever-increasing role in our daily life. Therefore, it is important to understand the potential risks involved with these devices. In this work we demonstrate the thermal runaway characteristics of three types of commercially available lithium-ion batteries with the format 18650. The lithium-ion batteries were deliberately driven into thermal runaway by overheating under controlled conditions. Cell temperatures up to 850 °C and a gas release of up to 0.27 mol were measured. The main gas components were quantified with gas-chromatography. The safety of lithium-ion batteries is determined by their composition, size, energy content, design and quality. This work investigated the influence of different cathode-material chemistry on the safety of commercial graphite-based 18650 cells. The active cathode materials of the three tested cell types were (a) $LiFePO_4$, (b) $Li(Ni_{0.45}Mn_{0.45}Co_{0.10})O_2$ and (c) a blend of $LiCoO_2$ and $Li(Ni_{0.50}Mn_{0.25}Co_{0.25})O_2$.

Keywords Lithium-ion battery · Thermal runaway · Gas analysis

3.1 Introduction

Lithium-ion batteries have been commercially available since 1991 [12]. As of 2013, lithium-ion batteries are in wide use for portable electronics, such as cell phones and notebook computers. They are also gaining traction as a power source in electrified vehicles. Lithium-ion batteries have a high specific energy and favourable ageing characteristics compared to NiMH and lead acid batteries. However, there are

A. W. Golubkov(✉) · D. Fuchs
Virtual Vehicle Research Center, Graz, Austria
e-mail: andrej.golubkov@v2c2.at

D. Fuchs
e-mail: david.fuchs@v2c2.at

A. Thaler and D. Watzenig (eds.), *Automotive Battery Technology*,
Automotive Engineering: Simulation and Validation Methods,
DOI: 10.1007/978-3-319-02523-0_3, © The Author(s) 2014

concerns regarding the safety of lithium-ion batteries. Abuse conditions such as over-charge, over-discharge and internal short-circuits can lead to battery temperatures far beyond the manufacturer ratings. At a critical temperature, a chain of exother-mic reactions can be triggered. The reactions lead to a further temperature increase, which in turn accelerates the reaction kinetics. This catastrophic self-accelerated degradation of the lithium-ion battery is called thermal runaway [17].

During thermal runaway, temperatures as high as 900 °C can be reached [6], and the battery can release a significant amount of burnable and (if inhaled in high concen-trations) toxic gas [13]. To quantify possible hazards of exothermic lithium-ion bat-tery over-temperature reactions, tests with complete batteries should be performed. Such experiments were undertaken with commercial lithium-ion batteries produced for consumer electronics [2, 4, 6–8, 10, 13, 15, 16] and with lithium-ion batteries fabricated in the laboratory [1, 3, 5, 11, 14].

This work investigated the thermal stability of three types of commercially avail-able lithium-ion batteries for consumer electronics. Particular attention was given to (1) the dynamics of the thermal responses of the cells, (2) the maximum tempera-tures reached, (3) the amount of gases produced and (4) to the production rates of the gases. To further assess the hazard potential of the released gases, samples were taken and analysed with a gas chromatography system.

3.2 Experimental

3.2.1 Brief Description of the Test Rig

To carry out unrestricted thermal-runaway experiments, a custom-designed test stand was built (Fig. 3.1). The main component of the test rig is a heatable reactor with electric feedthroughs for the temperature measurement and the inner sample heat-ing. The reactor has gas feedthroughs that connect it to an inert gas flask, to a gas sampling station and to a cold trap with an attached vacuum pump. The pressure inside the reactor is recorded by a pressure transmitter. The whole structure is hosted inside a fume hood to prevent any escaping of gases and electrolyte vapours. A removable sample holder is placed inside the reactor. The sample holder consists of a metal structure, which houses a heating sleeve and the thermocouples. A lithium-ion battery with the dimensions 18650 (cylindrical geometry with $d = 18$ mm and $l = 65$ mm) can be fitted into the centre of the heating sleeve. The inside wall of the heating sleeve is thermally insulated. The role of the thermal insulation layer is to provide the thermal connection between the heating sleeve and the sample. Due to the low thermal conductivity of the insulation layer, a thermal runaway reaction can proceed in adiabatic-like conditions. Ten thermocouples measure the temperature at different positions inside the reactor: three thermocouples are directly attached to the sample housing, three thermocouples are attached to the heating sleeve and four thermocouples measure the gas temperature inside the reactor.

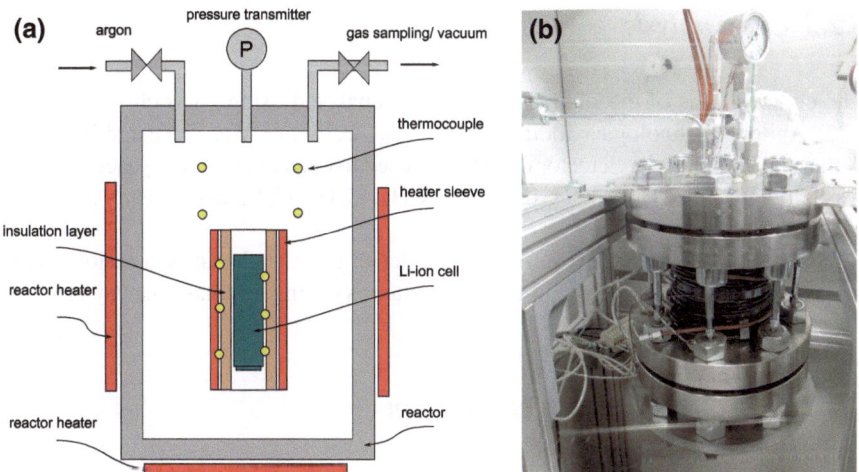

Fig. 3.1   **a** The reactor and its principal elements. **b** The reactor is the main component of the test stand

3.2.2 Testing Method

Initially, the sample battery is CC/CV charged to the respective cut-off voltage. Then, the plastic envelope is removed from the cell and the cell mass and cell voltage are recorded. Three thermocouples are welded to the cell housing, and the whole package is inserted into the heating sleeve of the sample holder. The sample holder is placed inside the reactor. The reactor is evacuated and flushed with argon gas twice. The heaters are set to constant power, and the pressure and temperature signals are recorded. In order to trace fast temperature and pressure changes, each signal is recorded with a high sampling rate of 5000 samples per second.

When a critical temperature is reached, the cell goes into rapid thermal runaway: it produces gas and heat. During the thermal runaway, the temperature of the cell increases by several hundred degree Celsius in a few seconds. After the thermal-runaway event, the cell cools down slowly. Gas samples are taken and analysed with the gas chromatograph. In the next step, the vacuum pump is switched on, and the cooling trap is filled with liquid nitrogen. The gas is carefully released through the cooling trap and the vacuum-pump into the fume hood. The reactor and the gas tubes between the reactor and the cooling trap are heated above 130 °C to avoid gas condensation.

By following this procedure, most liquid residue in the reactor is passed from the reactor to the cooling trap. The liquid residue can be easily removed from the cooling trap before the next experiment run.

3.2.3 Gas Analysis

The compositions of the sampled gases were analysed using a gas chromatograph (GC, Agilent Technologies 3000 Micro GC, two columns, Mol Sieve and PLOTU). A thermal conductivity detector (TCD) was used to detect permanent gases. The GC was calibrated for H_2, O_2, N_2, CO, CO_2, CH_4, C_2H_2, C_2H_4 and C_2H_6. Ar and He were used as carrier gases.

Note, that the current test set-up cannot detect HF, which can be a major source of toxicity of gas released by lithium-ion batteries during thermal runaway [13].

3.2.4 Cell-Components Identification

In order to identify the components of each cell species, several cells were disassembled: the cells were discharged to 2.0 V, and the cell casings were then carefully removed without causing short circuits. The exposed jelly rolls were subject to several tests.

For electrolyte identification, the jelly rolls were immersed in flasks with CH_2Cl_2 solution immediately after casing removal. The solutions were then analysed using a gas-chromatography—mass spectrometry system (GC-MS: Agilent 7890 and MS 5975MSD) with the ChemStation software and the NIST spectrum library. To analyse the solid materials of the cells, the extracted jelly rolls were separated into the anode, cathode and separator layers. After drying in a chemical fume hood, anode and cathode-foil samples were taken for identification of the electrochemically active materials. Microwave-assisted sample digestion followed by inductively coupled plasma optical emission spectrometry (ICP-OES, Ciros Vision EOP, Spectro, Germany) was used to obtain the gross atomic compositions of the cathode active masses. A scanning-electron microscope with energy-dispersive X-ray spectroscopy (SEM/EDX: Zeiss Ultra55 and EDAX Pegasus EDX) was used to confirm the ICP-OES results for the compositions of the cathodes and to validate the anode materials. For the mass-split calculation, the following procedure was followed for each cell type: Positive and negative electrode samples were extracted from the jelly roll. The samples were rinsed with diethyl carbonate (DEC) and then dried again, in order to remove the remaining electrolyte residues from the active materials. The samples were weighed, and the geometries of the electrode foils were recorded, so that the mass split could be calculated. The amount of electrolyte was estimated as the mass difference between the initial cell mass and the calculated dry mass for each cell. The thickness of the active material layers on the electrode substrates was extracted from SEM images. The thicknesses of the aluminium and copper substrates were calculated from the measured area density. The thickness of the separator foils was measured with a micrometer.

3.2.5 Lithium-Ion Cells

18650 consumer cells with three types of chemistry were purchased for the experiments. The cells were produced by three well-known companies. For simplicity, the samples will be referred to as LFP, NMC and LCO/NMC cells, in order to reflect their respective cathode material. Despite the simple naming scheme, please note that the cells do not differ in the types of their cathode material alone. Naturally, they also have different layer geometries (Table 3.2) and different ratios of their component masses (Fig. 3.2), and there are differences in the composition of the active masses as well (Table 3.1).

- The LCO/NMC cell had a blended cathode with two types of electrochemically active particles $LiCoO_2$ and $Li(Ni_{0.50}Mn_{0.25}Co_{0.25})O_2$. A clean cut through the sample was done with a focused ion beam (FIB). Subsequently, EDX measurements of the bulk materials of individual cathode particles were performed. The ratio of LCO and NMC layered oxide particles was estimated by comparing the SEM-EDX and ICP-OES results. The resulting ratio of LCO and NMC was LCO:NMC = (66:34) with 5 % uncertainty. The cells with LCO/NMC blended cathodes are a compromise to achieve high rate capability of LCO material and to maintain acceptable safety and high capacity of the NMC material [9]. The average voltage of this cell was ~3.6 V.
- The NMC cell had a $Li(Ni_{0.45}Mn_{0.45}Co_{0.10})O_2$ layered oxide cathode. The properties of the NMC mixed oxide cathodes depended on the ratios of nickel, manganese and cobalt material. In general, NMC cells have an average voltage of ~3.6 V and high specific capacity [18] .
- The LFP cell had a $LiFePO_4$ cathode with olivine structure. This cathode type is known for featuring good safety characteristics. Commercial $LiFePO_4$ cathode material for high power lithium-ion batteries consists of carbon-coated $LiFePO_4$ nano-scale particles. The cathode material is readily available and non-hazardous. Commercially available LFP cells have a lower operating voltage (~3.3 V) than cells with LCO and NMC cathodes [18].

The active anode materials consisted only of carbonaceous material for all cells, as verified by SEM/EDX. The exact types of graphite materials could not be identified.

3.2.6 Electrical Characterisation

An electrical characterisation of the cells was done with a BaSyTec CTS cell test system. In the first step, the cells were discharged to their respective minimum voltage. In the second step, the cells were charged using a pulse-pause protocol, until the voltage of the cells stayed above their respective maximum voltage during a pause. The current pulses were set to 100 mA and 30 s. The duration of the pauses was set to 50 s. The open-circuit voltage (OCV) at the end of each pause and the charge

Table 3.1 Overview of the three cells species used in the experiments. All ratios in this table are given as mol ratios. The electrolyte solvents are dimethyl carbonate (DMC), ethyl methyl carbonate (EMC), ethylene carbonate (EC) and propylene carbonate (PC)

		LCO/NMC	NMC	LFP
Cell mass	g	44.3	43.0	38.8
Capacity	Ah	2.6	1.5	1.1
Minimal voltage	V	3.0	3.0	2.5
Maximal voltage	V	4.2	4.1	3.5
Solvents (DMC:EMC:EC:PC)		6:2:1:0	7:1:1:1	4:2:3:1
Cathode material		2/3 LCO + 1/3 NMC(211)	NMC(992)	$LiFePO_4$
Anode material		Graphite	Graphite	Graphite

Table 3.2 Mass (m), area (A), thickness (d) and volume (V) of the main components of the three cell species. The geometrical volume of a standard 18650 cell is $16.5\,cm^3$

	LCO/NMC				NMC				LFP			
	m	A	d	V	m	A	d	V	m	A	d	V
	g	cm^2	μm	cm^3	g	cm^2	μm	cm^3	g	cm^2	μm	cm^3
Separator	1.2	942	19	1.8	1.4	944	23	2.2	1.2	940	20	1.9
Cathode Al foil	1.7	403	16	0.6	3.1	389	30	1.1	2.1	396	19	0.7
Cathode active material	18.3	715	91	6.5	11.3	654	67	4.4	9.7	793	70	5.5
Anode Cu foil	2.9	402	8	0.3	7.5	418	20	0.8	3.9	396	17	0.7
Anode active material	8.1	739	81	6.0	6.2	695	60	4.2	5.2	793	50	4.0
Electrolyte	4.6				4.4				6.4			
Housing	7.5				9.2				10.5			
Sum	44.3			15.2	43.1			12.7	39.0			12.8

capacity were recorded (Fig. 3.3). For the NMC cell, the cell manufacturer did not provide the voltage ratings. For safety reasons, 4.1 V was selected as the maximum voltage.

3.3 Results and Discussion

3.3.1 Typical Course of a Thermal Runaway Experiment

In order to illustrate the events during the heat-up process and the thermal runaway itself, one experiment with a NMC cell is described here in detail:

The NMC sample cell was prepared as described above. At the start of the test, the cell heater sleeve was set to constant heating power. The sample was slowly heated, starting at 25 °C, with a heat-rate of ~2 °C/min. After reaching 220 °C, the cell went into rapid thermal runaway. The cell temperature rose from 220 to 687 °C in a few seconds. When the exothermic reaction ended, the cell cooled down slowly (Fig. 3.4a).

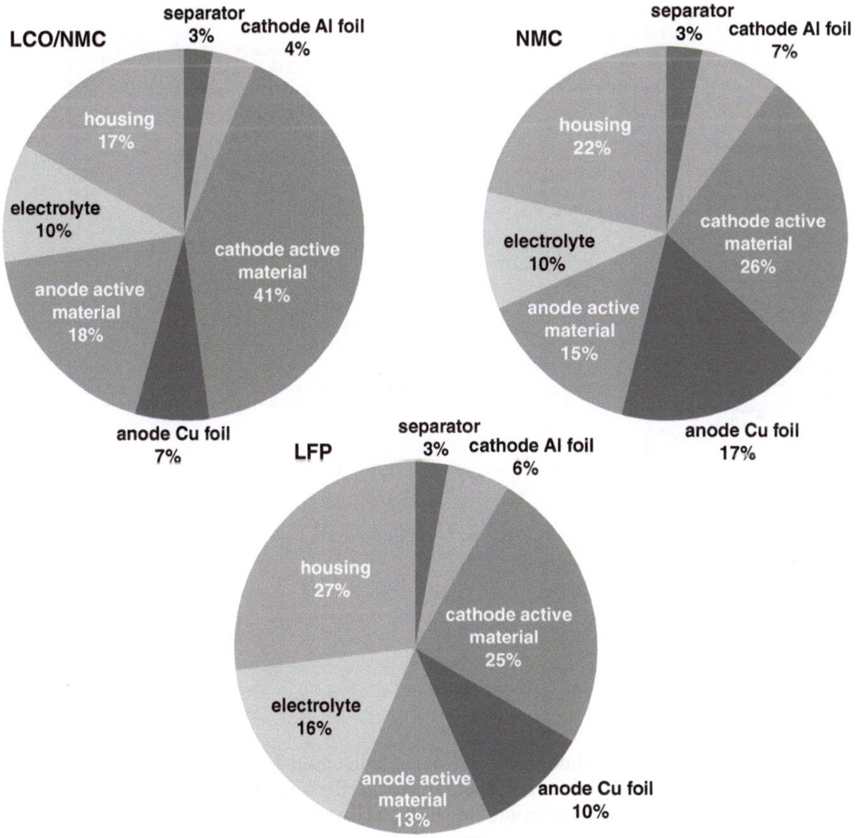

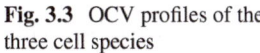

Fig. 3.2 Mass split (mass %) of the main components of the three cell species

Fig. 3.3 OCV profiles of the three cell species

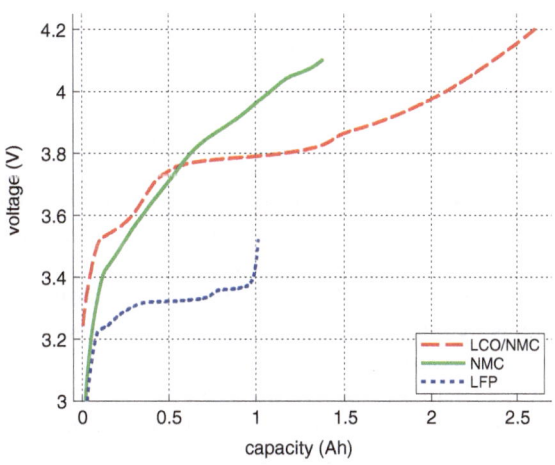

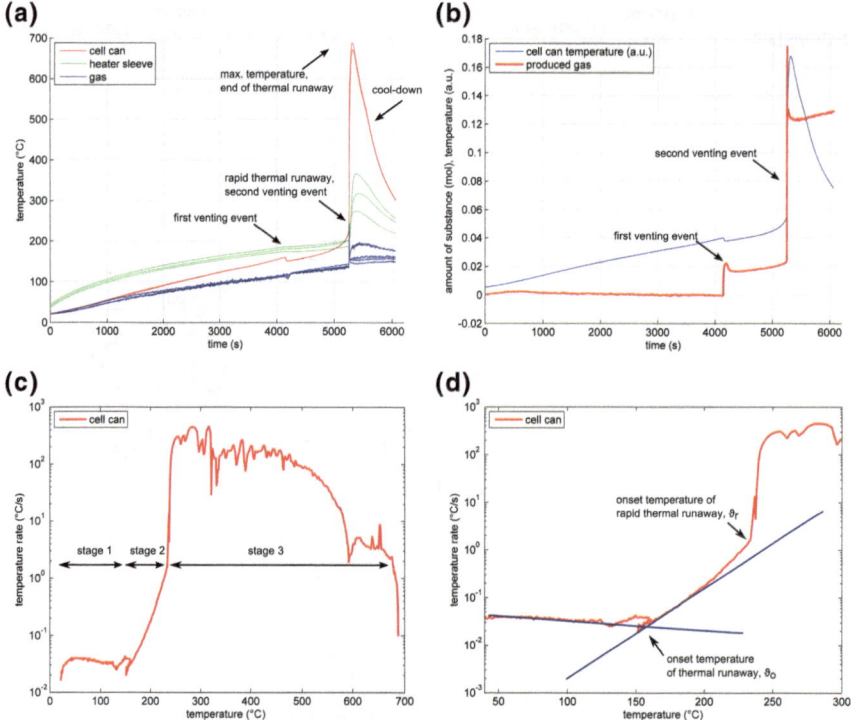

Fig. 3.4 **a** Temperature versus time plot of all temperature sensors in the pressure vessel. The whole duration of the experiment is shown. **b** Amount of produced gas versus time plot. Cell temperature is shown in arbitrary units. **c** Temperature rate of the cell versus cell temperature. Overview of a whole experiment duration. **d** Temperature rate of the cell versus cell temperature. The straight lines are fitted to the heat-up stage and to the quasi-exponential stage. The intersection of the two lines marks the onset point θ_o of the thermal runaway reaction. A sharp increase in the temperature rate marks the onset of the rapid thermal runaway θ_r

The amount of gas produced inside the pressure vessel was calculated by applying the ideal gas law:

$$n = \frac{pV}{R\theta_{gas}} - n_0 \qquad (3.1)$$

where p is the recorded pressure in the reactor, $V = 0.0027\,\mathrm{m}^3$ is the reactor volume, R is the gas constant, θ_{gas} is the recorded gas temperature in the reactor (in K), and n_0 is the initial amount of gas in the reactor at the start of the experiment.

At 160 °C, the safety vent device of the battery housing opened, and 0.02 mol of gas were released by the cell. The cell cooled down by 10 °C during the release process because of the Joule-Thomson effect. The vent opening was then probably clogged until, at 230 °C, concurrent with the rapid thermal runaway, the cell vented for a second time. The second venting was the major venting: an additional 0.15 mol of vent gas were produced (Fig. 3.4b).

Note that the amount of gas in the reactor decreased shortly after venting. This effect can be explained by the condensation of gas at the reactor walls. Since the reactor walls had a lower temperature ($\sim$150 °C) than the cell in full thermal runaway (up to 687 °C), the walls could act as a gas sink.

In order to visualise subtle changes in thermal behaviour of the cell during the experiment, rate diagrams are utilized. Contrary to a common temperature versus time diagram (θ vs. t), the temperature rate is plotted versus temperature ($d\theta/dt$ vs. θ) in a rate diagram. This type of diagram is often used to visualise accelerating rate calorimetry (ARC) results as well. Three distinct experiment stages can be seen in the rate diagram for the NMC cell (Fig. 3.4c):

1. Heat-up stage ($\theta < \theta_o$): In the temperature range from room temperature to θ_o at $\sim$170 °C, the cell generated no heat. The heater sleeve was the only heat source in this phase. The negative peak at 130 °C is associated with endothermic separator melting. (It is analogous to a negative endothermic peek in a differential scanning calorimetry (DSC) diagram during the phase change of a sample.) The temperature θ_o at which a cell starts to generate heat is commonly called the onset temperature of the thermal runaway.
2. Quasi-exponential heating stage ($\theta_o < \theta < \theta_r$): At temperatures higher than θ_o, the battery became a heat source. Between 170 and 220 °C, the temperature rate increase followed a nearly straight line in the logarithmic plot (Fig. 3.4d). At 220 °C, a sharp increase in temperature rate marked the end of the quasi-exponential heating stage.
3. Rapid thermal runaway stage ($\theta_r < \theta < \theta_m$): At 220 °C, θ/dt increased sharply and initiated the rapid thermal runaway. The transition to thermal runaway was accompanied by a venting event. The thermal runaway ended when all reactants had been consumed. At this point, the maximum temperature $\theta_m = 687$ °C was reached.

It is difficult to pinpoint the exact transition between stage 1 and 2. Several endothermic events often occurred near the onset temperature θ_o: the separator melt temperature was 130 °C, the cell safety vent usually opened at 160 °C and some material was released from the cell, causing a slight cool-down due to the Joule-Thomson effect. Thus, the exact value of θ_o can be obscured by the intermediate cell cool-down. To keep it simple, θ_o was defined as the point at which the heating-rate curve switches from constant to quasi-exponential rising. One line is fitted to the heat-up part and one line to the quasi-exponential part of the rate curve in the logarithmic rate plot. The onset temperature θ_o can be further defined as the temperature at which the two lines cross (Fig. 3.4d).

3.3.2 Thermal-Runaway Experiments

At least three thermal-runaway experiments were conducted with each of the three cell species. A temperature profile overview of all experiments is shown in Fig. 3.5a.

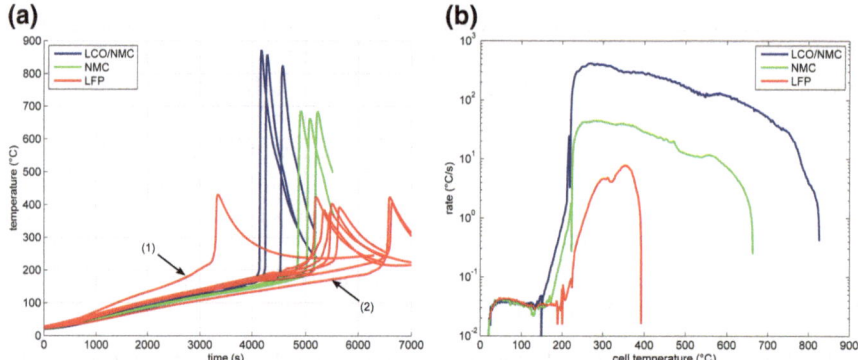

Fig. 3.5 **a** Overview of the time-temperature profiles for the cells tested. Data for the whole experiment durations and for the whole experiment sets is shown. For the sake of completeness, one LFP test with a higher (1) and one with a lower (2) heating rate of the heater sleeve are included. **b** Temperature rates from three representative experiments

Each species had its unique thermal-runaway characteristics. The high capacity, cobalt rich LCO/NMC cells reached the highest θ_m at $(853 \pm 24)\,°C$ during thermal runaway. The cobalt poor NMC cells had a lower θ_m of $(678 \pm 13)\,°C$. The LFP cells showed a less pronounced thermal runaway and reached a moderate θ_m of $(404 \pm 23)\,°C$. The temperature curves showed small variations from sample to sample. It is likely that the variations were caused by different burst times of the rupture plates, which, together with subtle effects of venting, Joule-Thomson cool-down and clogging of the vent openings, influence the thermal-runaway reaction-pathways.

For the sake of completeness, two additional LFP experiments with different heater-sleeve heating-rates (1.5 and 3.5 °C/min) were also included in the analysis (Fig. 3.5a). The thermal runaway characteristics of the LFP cell (θ_r, θ_m and n) did not depend on the heater-sleeve heating rate in the given heat-rate range. The two additional experiments contributed to the mean values in Table 3.3 and Fig. 3.6. For clarity, only one representative curve for each cell species is shown in the following graphs.

Each cell species had distinctive kinetic thermal-runaway characteristics (Table 3.3 and Fig. 3.5b). Of the three specimen, the LCO/NMC cell showed the lowest θ_o and θ_r, hence the LCO/NMC cell was the cell most vulnerable to over-heating conditions. For the NMC cell, θ_o and θ_r were shifted to higher temperatures. Transition temperatures of the LFP specimen were noticeably higher than those of both metal/oxide cells (LCO/NMC and NMC). The LFP cell was able to withstand the highest temperature before going into thermal runaway.

Both metal oxide cells showed the three stages described above (heat-up, quasi exponential heating, rapid thermal runaway). In contrast, the thermal runaway profile of the LFP cell lacked a distinct quasi-exponential stage. For the LFP cell, it was difficult to find a clear distinction between θ_o and θ_r. Therefore, θ_r is not given for the LFP species.

Table 3.3 Characteristic temperatures and venting parameters in the thermal-runaway experiments. Here, θ_o is the onset temperature, θ_r is the transition temperature into rapid thermal runaway, θ_m is the maximum recorded temperature, n is the total amount of gas produced as measured in the reactor at a reactor temperature of 150 °C, and Δt is the typical venting duration

		LCO/NMC	NMC	LFP
θ_o	°C	149 ± 2	168 ± 1	195 ± 8
θ_r	°C	208 ± 2	223 ± 3	–
θ_m	°C	853 ± 24	678 ± 13	404 ± 23
n	mmol	265 ± 44	149 ± 24	50 ± 4
Δt	s	0.8	0.2	30.0

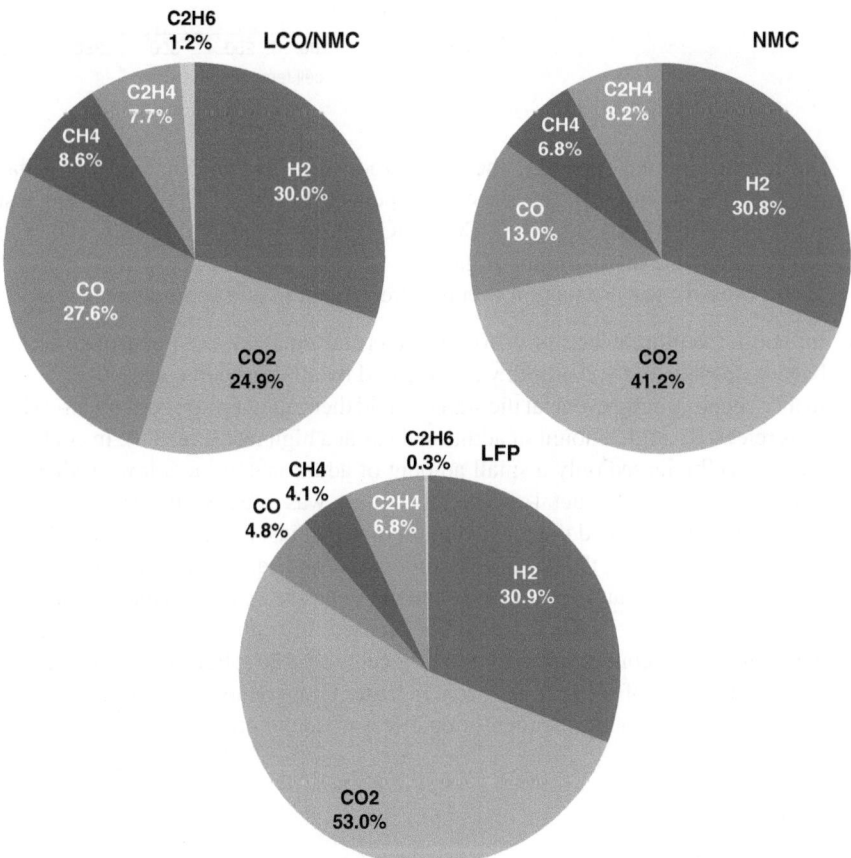

Fig. 3.6 Detected components of the produced gases (mol %)

Fig. 3.7 Temperature-
ventgas profiles. Note that
the *x*-axis is trimmed to the
relevant temperature range

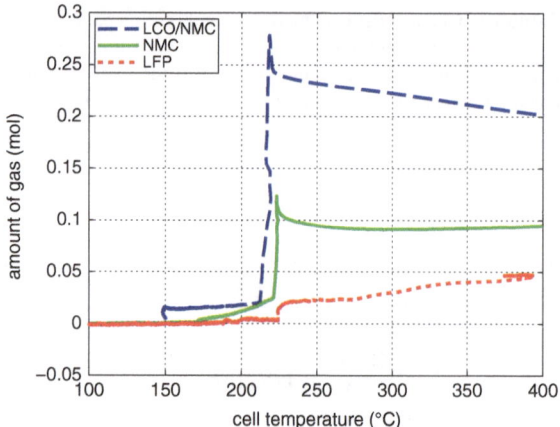

During the thermal runaway, the cells produced a significant amount of gas
(Table 3.3). The amount of gas strongly depended on the cell type. The highest
amount of gas was released by the LCO/NMC cell, followed by the NMC cell. The
LFP cell yielded the least amount of gas.

Two successive gas production events were evident in all experiments (Fig. 3.7):

1. In the first venting event, prior to rapid thermal runaway, the burst plate of the
   battery opened, and ~20 mmol were released by all three cell types.
2. In the second venting event, at the start of rapid thermal runaway, both metal-oxide
   cells released a high amount of additional gas at a high rate (Fig. 3.8). In contrast,
   the LFP cell released only a small amount of additional gas at a low production
   rate. In the case of the metal-oxide cells the gas was released in very short time.
   The NMC cell produced the main amount of gas in just 0.2 s, and the LCO/NMC
   in 0.8 s. After release, the hot gas was not in thermal equilibrium with the cooler
   walls of the reactor, and therefore the amount of gas decreased, as the released gas
   came into contact with the walls and condensed. In contrast, the gas production
   duration of the second venting for the LFP cell was ~30 s. Because of the gradual
   release, the gases of the LFP cell were in better temperature equilibrium with the
   reactor walls and the gas condensation effect was not noticeable.

3.3.3 Gas Analysis

At least one gas analysis was performed for each cell species. Each cell type showed
a unique gas composition footprint (Fig. 3.6). The main components were H_2 and
CO_2. Both metal-oxide cells produced a significant amount of CO. Additionally,
smaller fractions of CH_4, C_2H_4 and C_2H_6 were identified. As mentioned before, HF
was not measured.

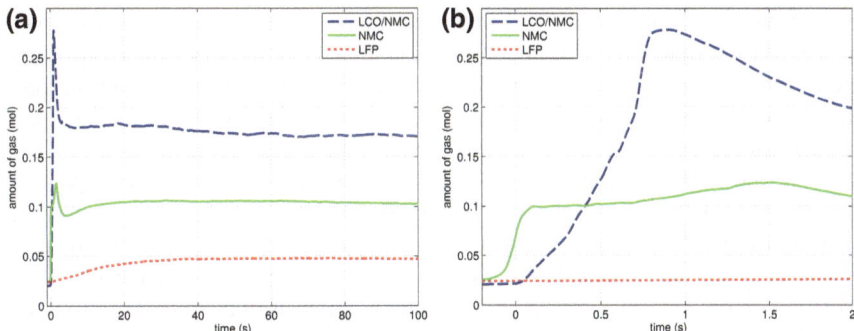

Fig. 3.8 Time-ventgas profiles. Note: to make the curves comparable, each curve was moved on the time axis, so that the second venting event starts at time zero. **a** The first 100 s and **b** the first 2 s of the second venting events are shown

Most components of the gases are flammable. The gases can be toxic due to the presence of CO.

3.4 Conclusion and Outlook

Three types of consumer lithium-ion batteries with the format 18650 with different cathode materials were evaluated in thermal runaway tests. The cells were brought into thermal runaway by external heating. All tests were performed in a pressure-tight reactor in an argon atmosphere. In agreement with literature [4], the cell containing LFP showed the best safety characteristics. The LFP cell had the highest onset temperature (~195 °C), the smallest temperature increase during the thermal runaway (~210 °C), the lowest amount of produced gas (~50 mmol) and the lowest percentage of toxic CO in the gas (~4 %). Unfortunately, it was also the cell with the lowest working voltage (3.3 V) and the lowest energy content (3.5 Wh). Batteries with higher energy content (5.5 and 9.6 Wh) performed worse in safety tests. The onset temperature shifted down to ~170 and ~150 °C, the temperature increase during thermal runaway rose to ~500 and ~700 °C, the amount of gas released was ~150 and ~270 mmol, and significant percentages of CO (13 and 28 %) were found for the NMC and NMC/LCO cells, respectively.

All cells released high amounts of H_2 and hydrocarbons. These gases are highly flammable. Even though the gas could not burn in the inert atmosphere inside the reactor, the surface of the high-energy cells reached temperatures of up to 850 °C during the experiments.

Modern devices are equipped with battery temperature and voltage monitoring. If a state beyond specification is detected, the devices shut down automatically to prevent battery abuse [18]. If system shut-down can not prevent a thermal runaway in all cases, data in this work may be a valuable source for the specification of a robust energy-storage system which can withstand conceivable abuse events.

To reduce possible damage from thermal-runaway events in consumer devices, we suggest the following design optimization targets: (1) increase the temperature endurance and heat absorption capability of used materials; (2) minimize heat propagation to neighbouring burnable elements; (3) minimize gas ignition probability (e.g. mechanical separation of electric components from the gas release position). This work has shown that the kinetics of the thermal-runaway process strongly depend on the energy content of the lithium-ion battery. Future work will focus on the thermal runaway triggered by over-heating at different states of charge (SoC) and the thermal runaway caused by overcharge. Emphasis will be given to assessment of HF gas evolution, to gas analysis with GC-MS, and to the analysis of the liquid residues that are collected in the cooling trap.

Acknowledgments The authors would like to acknowledge the financial support of the "COMET K2—Competence Centres for Excellent Technologies Programme" of the Austrian Federal Ministry for Transport, Innovation and Technology (BMVIT), the Austrian Federal Ministry of Economy, Family and Youth (BMWFJ), the Austrian Research Promotion Agency (FFG), the Province of Styria and the Styrian Business Promotion Agency (SFG).

We would furthermore like to express our thanks to our supporting scientific project partners, namely Graz Centre for Electron Microscopy and the Graz University of Technology, Institute of Chemical Engineering and Environmental Technology.

The chapter 'Thermal Runaway—Causes and Consequences on Cell Level' is derived from the original article 'Thermal-runaway experiments on consumer Li-Ion batteries with metal-oxide and olivin-type cathodes' RSC Advances, 4(7), 3633. doi:10.1039/c3ra45748f, (2014), by A. W. Golubkov, D. Fuchs, J. Wagner, H. Wiltsche, C. Stangl, F. Gisela, G. Voitic, A. Thaler, and V. Hacker—Reproduction by permission of The Royal Society of Chemistry.

References

1. Abraham D, Roth EP, Kostecki R, McCarthy K, MacLaren S, Doughty D (2006) Diagnostic examination of thermally abused high-power lithium-ion cells. J Power Sources 161(1):648–657. doi:10.1016/j.jpowsour.2006.04.088, http://linkinghub.elsevier.com/retrieve/pii/S0378775306006768

2. Belov D, Yang MH (2008) Investigation of the kinetic mechanism in overcharge process for Li-ion battery. Solid State Ionics 179(27–32):1816–1821. doi:10.1016/j.ssi.2008.04.031, http://linkinghub.elsevier.com/retrieve/pii/S0167273808003858

3. Chen Z, Qin Y, Ren Y, Lu W, Orendorff C, Roth EP, Amine K (2011) Multi-scale study of thermal stability of lithiated graphite. Energy Environ Sci 4(10):4023. doi:10.1039/c1ee01786a, http://xlink.rsc.org/?DOI=c1ee01786a

4. Doughty D, Roth EP (2012) A general discussion of Li ion battery safety. Electrochem Soc Interface 21(2):37–44. http://www.scopus.com/inward/record.url?eid=2-s2.084867753898&partnerID=40&md5=19382decb891d60f28ef1049fca727ea

5. Doughty DH, Roth EP, Crafts CC, Nagasubramanian G, Henriksen G, Amine K (2005) Effects of additives on thermal stability of Li ion cells. J Power Sources 146(1–2):116–120. doi:10.1016/j.jpowsour.2005.03.170, http://linkinghub.elsevier.com/retrieve/pii/S0378775305005057

6. Jhu CY, Wang YW, Shu CM, Chang JC, Wu HC (2011a) Thermal explosion hazards on 18650 lithium ion batteries with a VSP2 adiabatic calorimeter. J Hazard Mater 192(1):99–107. doi:10.1016/j.jhazmat.2011.04.097, http://www.ncbi.nlm.nih.gov/pubmed/21612866

7. Jhu CY, Wang YW, Wen CY, Chiang CC, Shu CM (2011b) Self-reactive rating of thermal run-away hazards on 18650 lithium-ion batteries. J Therm Anal Calorim 106(1):159–163. doi:10.1007/s10973-011-1452-6, http://link.springer.com/10.1007/s10973-011-1452-6

8. Jhu CY, Wang YW, Wen CY, Shu CM (2012) Thermal runaway potential of LiCoO2 and Li(Ni1/3Co1/3Mn1/3)O2 batteries determined with adiabatic calorimetry methodology. Appl Energy 100:127–131. doi:10.1016/j.apenergy.2012.05.064, http://linkinghub.elsevier.com/retrieve/pii/S0306261912004655

9. Lee KS, Myung ST, Kim DW, Sun YK (2011) AlF3-coated LiCoO2 and Li[Ni1/3Co1/3Mn1/3]O2 blend composite cathode for lithium ion batteries. J Power Sources 196(16):6974–6977. doi:10.1016/j.jpowsour.2010.11.014, http://linkinghub.elsevier.com/retrieve/pii/S0378775310019208

10. Maleki H, Deng G, Anani A, Howard J (1999) Thermal stability studies of Li-ion cells and components. J Electrochem Soc 146(9):3224. doi:10.1149/1.1392458, http://link.aip.org/link/?JES/146/3224/1&Agg=doi

11. Nagasubramanian G, Orendorff CJ (2011) Hydrofluoroether electrolytes for lithium-ion batteries: reduced gas decomposition and nonflammable. J Power Sources 196(20):8604–8609. doi:10.1016/j.jpowsour.2011.05.078, http://linkinghub.elsevier.com/retrieve/pii/S0378775311011049

12. Nagaura T, Tozawa K (1990) Lithium ion rechargeable battery. Prog Batteries Sol Cells 9:209

13. Ribière P, Grugeon S, Morcrette M, Boyanov S, Laruelle S, Marlair G (2012) Investigation on the fire-induced hazards of Li-ion battery cells by fire calorimetry. Energy Environ Sci 5(1):5271. doi:10.1039/c1ee02218k, http://xlink.rsc.org/?DOI=c1ee02218k

14. Roth EP, Orendorff CJ (2012) How electrolytes influence battery safety. Electrochem Soc Interface 21(2):45–49. http://www.scopus.com/inward/record.url?eid=2-s2.084867822714&partnerID=40&md5=7ce53080e26e92d559c78118e5cd0e87

15. Tobishima S, Yamaki J (1999) A consideration of lithium cell safety. J Power Sources 81–82:882–886. doi:10.1016/S0378-7753(98)00240-7, http://linkinghub.elsevier.com/retrieve/pii/S0378775398002407

16. Wen CY, Jhu CY, Wang YW, Chiang CC, Shu CM (2012a) Thermal runaway features of 18650 lithium-ion batteries for LiFePO4 cathode material by DSC and VSP2. J Therm Anal Calorim 109(3):1297–1302. doi:10.1007/s10973-012-2573-2, http://link.springer.com/10.1007/s10973-012-2573-2

17. Wen J, Yu Y, Chen C (2012b) A review on lithium-ion batteries safety issues: existing problems and possible solutions. Mater Express 2(3):197–212. doi:10.1166/mex.2012.1075, http://openurl.ingenta.com/content/xref?genre=article&issn=2158-5849&volume=2&issue=3&spage=197

18. Zhang ZJ, Ramadass P (2012) Encyclopedia of sustainability science and technology. Springer, New York. doi:10.1007/978-1-4419-0851-3, http://link.springer.com/10.1007/978-1-4419-0851-3

Chapter 4
Application-Related Battery Modelling: From Empirical to Mechanistic Approaches

Franz Pichler and Martin Cifrain

Abstract Mathematical modelling and simulation has been an essential part of battery research and development ever since. Depending on the particular, objective several different approaches are feasible, each of which provides specific advantages, e.g. calculation speed or deep mechanistic insight. This chapter presents an overview of common battery model approaches and introduces the multi-scaling technique for the simulation of larger battery units.

Keywords Battery · Physics-based modelling · Equivalent circuit modelling · Multi scaling

4.1 Introduction

Battery models are used to describe and predict the performance of batteries. A validated model at hand usually reduces the cost of experiments significantly. In principle, the models consist of one or more mathematical equations. The complexity of these models varies from simple empirical relationships based on one parameter to highly sophisticated mechanistic 3-dimensional models, comprising many partial differential equations with dozens of parameters, some of them anisotropic. It is obvious that the calculation effort greatly increases with model complexity. Therefore a careful selection of the model level is essential in order to meet the given requirements.

An erratum to this chapter is available at 10.1007/978-3-319-0253-0_8 .

F. Pichler (✉) · M. Cifrain
Virtual Vehicle Research Center, Graz, Austria
e-mail: franz.pichler@v2c2.at

M. Cifrain
e-mail: martin.cifrain@v2c2.at

A. Thaler and D. Watzenig (eds.), *Automotive Battery Technology*,
Automotive Engineering: Simulation and Validation Methods,
DOI: 10.1007/978-3-319-02523-0_4, © The Author(s) 2014

Fig. 4.1 State-of-charge
determination from specific
gravity of the electrolyte
in a lead-acid battery. *Dots*
Measured values [11], *lines*
empirical model approaches
(see text)

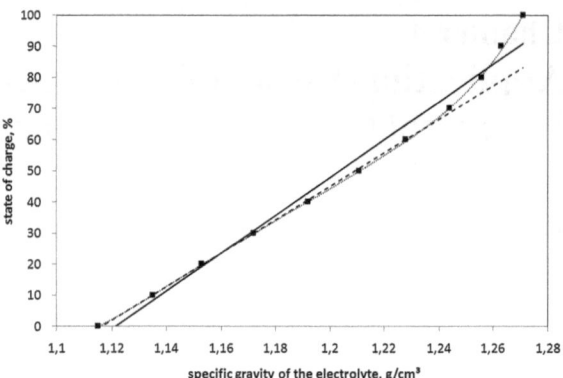

4.2 Empirical Models

The simplest—and earliest—battery models consist of just one equation, which
describes an observed relationship and usually does not take any given physical prop-
erty of the investigated battery into account. Such models are also called "black-box
models". A typical example is the calculation of the state of charge (SoC) of a lead-
acid battery from the specific gravity ρ of its electrolyte. Based on a measurement
(dots in Fig. 4.1 [11]), one can directly establish a best-fit line (continuous lines in
Figs. 4.1 and 4.2):

$$SOC = 606.84 \cdot \rho - 680.42. \tag{4.1}$$

The deviation plot (Fig. 4.2) shows that the deviation is—besides the point at 100 %
SoC—lower than 5 % (in absolute numbers); the model is fast and simple. However,
there is room for improvement. First, the valid range of the model can be reduced,
e.g. 0–70 % SoC. A linear approach would then result in (dashed lines in Figs. 4.1
and 4.2)

$$SOC = 537.31 \cdot \rho - 599.69. \tag{4.2}$$

This model gives much better results in the defined range (≤ 1 %), but fails over 80 %.
Hence, one can use this model if the SOC does not exceed 80 %. Second, the model
order can be increased, e.g. by a quadratic term:

$$SOC = 1565.5 \cdot \rho^2 - 3138 \cdot \rho + 1554.9. \tag{4.3}$$

The advantage of the increased complexity is a better approximation (dotted line in
Figs. 4.1 and 4.2) within the entire range.

Black-box models usually do not provide information about mechanisms inside
the cells, but they are fast and accurate within given ranges. The main intention of
showing this simple model comparison was to sensitize the reader to the important
topic of proper model selection. A model of the 6th order would have had a maximum

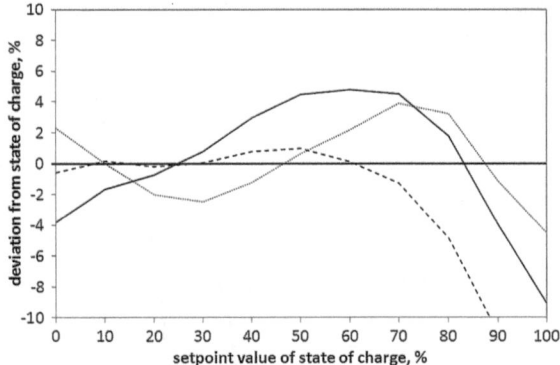

Fig. 4.2 Model deviations, lines related to Fig. 4.1

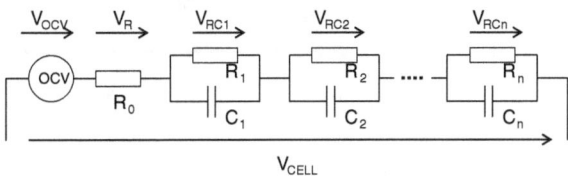

Fig. 4.3 A common equivalent circuit for modelling electric behaviour of battery cells

error of 1 %, but it is far from being useful in practice, as even the measurement method of the specific gravity is less accurate (error of the dots in Fig. 4.1). This holds true for all other more sophisticated approaches as well.

4.3 Equivalent Circuit Modelling

It has been observed that batteries behave somewhat like electronic circuitscomprising resistances and capacitors (see Electrochemical Impedance Spectroscopy (EIS) [5]), especially during dynamic loads. Consequently, it has become very common to model the current/voltage behaviour by equivalent circuits, using known relationships from electronics. These so-called RC models are usually fast, but again provide no insight into the battery. RC models can also be considered to be empirical, although some of the parameters can be attributed to the physical condition of the cells, e.g. two opposite plates (electrodes) as capacitor and the internal ohmic resistance to electrolyte conductivity.

A very common approach to an equivalent circuit is shown in Fig. 4.3. The first two parts in the figure are a voltage source that replicates the open circuit voltage of a battery and the inner resistance, which is responsible for the ohmic drop in voltage when switching on a current. The main components of this network are the RC circuits, each having the impedance

Fig. 4.4 Nyquist plot of
lithium-ion cell measured by
EIS (*dots*) and derived 2-RC
model (*continuous line*)

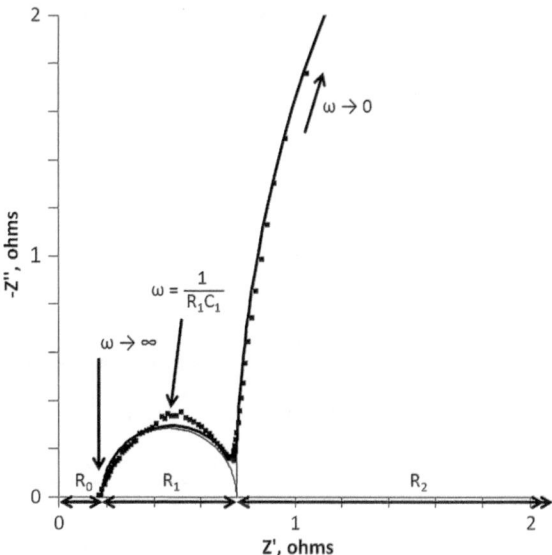

$$ Z_{RC} = \frac{R}{1 + \omega^2 R^2 C^2} - i \frac{\omega R^2 C}{1 + \omega^2 R^2 C^2} = Z' + iZ'', \qquad (4.4) $$

with ω ...angular frequency of an alternating current (AC) voltage that is applied to
the circuit and Z' and Z'' ...real and imaginary parts of the impedance. The angular
frequency is related to the AC frequency ν of the applied voltage by $\omega = 2\pi\nu$. In
Fig. 4.4 the *Nyquist plot* of a lithium-ion battery is shown (see Sect. 4.3.1). This type
of plot is widely used to show dynamic cell characteristics and shows the imaginary
part of the impedance over the real part for several angular frequencies. The lower
left part of the plot (thin solid line) shows a perfect semi-circle, which is exactly
the contribution of a single RC-circuit. Its highest point is at a frequency of exactly
$\omega = \frac{1}{RC}$ and its diameter is exactly R.

4.3.1 Parametrization

As shown above, one option for setting up an equivalent circuit model for a battery
is to simply use the *Nyquist plot* of a battery to fit the model parameters ($R_i, C_i, i = 1, \ldots, n$). Figure 4.4 (dots) shows an example of an EIS measurement [5] as a *Nyquist
plot*, whereby a lithium-ion cell was stimulated via an AC current of increasing
frequencies (approx. 2 Hz to 2 kHz). As every RC element delivers one semi-circle
(provided the capacitances are sufficiently different), one can see that there must be
at least two RC elements. The offset from zero on the x-axis is the ohmic internal
resistance (R_0). The small semi-circle is usually assigned to the electrode reaction.

The measured curve does not show a clean circle, which indicates that there is more than just one RC element behind it, but the fit still seems quite good. The large circle is assigned to transport processes. It can be modelled either via a large semi-circle or a 45°-straight line, called *Warburg Impedance*. While the location of the semi-circles along the x-axis is defined by the resistances, the depth of the valley between the circles is defined by the ratio of the capacitances (the larger the deeper). The absolute values of the capacitances can be calculated, as on the highest point of the circle the circular frequency equals $1/_{RC}$.

These parameters are highly dependent on several quantities, such as the temperature, the SoC of the battery and the current amplitude. Consequently, the described steps for parametrization have to be executed for many load points, thereby establishing multidimensional tables that give the RC parameter dependences on all the influence quantities considered.

Having assembled these look-up tables, the parameters can be adjusted dynamically using an interpolation scheme to obtain values for the actual load point in a simulation.

4.4 Mechanistic Models

Neither empiric nor RC models provide insight into the physical processes proceeding during cell operation. For studies of these processes in battery development and optimization, mechanistic models, also called electrochemical (EC) or physics-based models, are required. These models can easily grow to a high complexity comprising dozens of parameters, and many different approaches have been reported in the last decades. However, they all try to describe the coupled ongoing processes of the battery.

In most cases, an electrochemical cell consists of two different solid electrodes submerged in a liquid electrolyte (see Fig. 4.5). The electrodes form system-specific equilibrium electrode potentials, which can be measured as cell voltage. Furthermore, each electrode has the ability to provide a charge-transfer reaction (redox reaction) on its surface, which changes the charge-transport mechanism from electronic (in electrode) to ionic (in electrolyte) or the reverse. The electron current shows characteristics described by *Ohm's law* while the ionic current is diffusion controlled. Once connected, the cell is able to deliver a usable current, as long as active reducible and oxidizable material is present. A separator prevents the electrodes from internal short-circuiting.

The main aspects of mechanistic modelling are addressed below, as a detailed description would go beyond the scope of this introduction. For the same reason, boundary and initial conditions for the partial differential equations (in this and the next chapter (Sect. 4.5)) will only be given, when they play a key role in the modelling process. For more details on electro chemistry and mechanistic modelling see [2, 3, 7, 8].

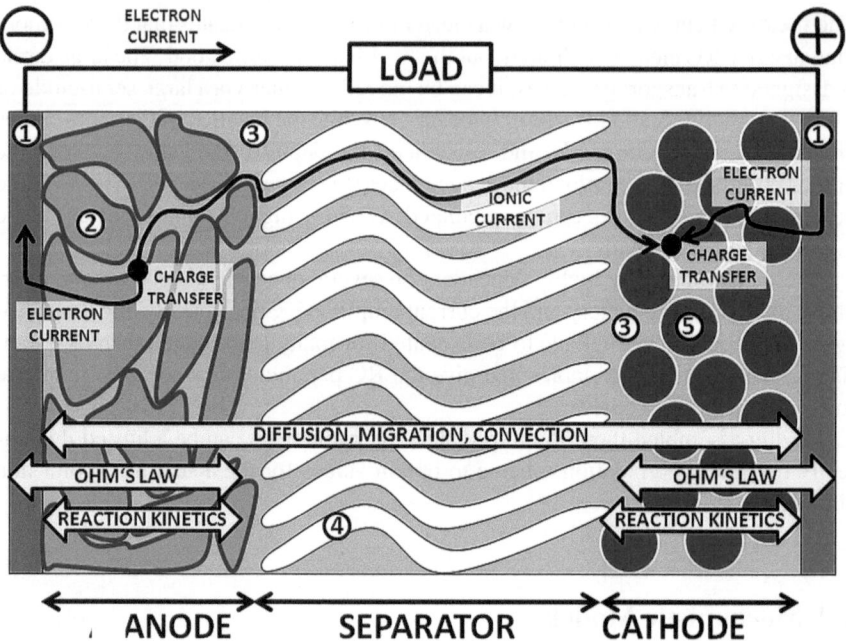

Fig. 4.5 Schematic view of a battery during discharge, comprising porous electrodes and positive ions transporting the charge. *1* current collectors, *2* anode particles, *3* electrolyte, *4* separator (e.g. porous plastics), *5* cathode particles

4.4.1 Charge Transfer

At the electrode-electrolyte interface, charge separation will occur forced by the electrochemical potential until an equilibrium condition is reached:

$$\text{reduced species} \rightleftharpoons \text{oxidized species} + \text{electron(s)},$$

$$\text{e.g. for zinc} \qquad Zn \rightleftharpoons Zn^{2+} + 2e^-. \tag{4.5}$$

In equilibrium, the amount of the forward reaction (from left to right) equals the amount of the backward reaction. While the electrons (e^-) stay in the electrode, the ions are dissolved by the electrolyte. However, due to electrostatic forces, the ions stay close to the interface to form an electric double layer, which acts like a capacitor. Hence, this electric charge separation builds up an electric potential difference E_0 between the electrode and the electrolyte, depending on the free energy ΔG of the system:

$$E^0 = -\frac{\Delta G}{nF}, \tag{4.6}$$

with F ...*Faraday constant* (≈ 96485 Cmol^{-1}) and n ...number of electrons in the reaction equation. This relationship is called *Faraday's Law*. The practically occurring potential difference E also depends on the concentration of the ions of the reduced and oxidized species, c_{RED} and c_{OX} respectively. If either species is solid, c equals 1 by definition (no dependence). The relation between E and c is the *Nernst Equation*:

$$E = E^0 - \frac{RT}{nF} \ln \left(\frac{\nu_{RED} c_{RED}}{\nu_{OX} c_{OX}} \right), \tag{4.7}$$

with R ...gas constant (≈ 8.314 Jmol^{-1} K^{-1}), T ...temperature (in K) and ν_{RED} and ν_{OX} the activity coefficients, which equal 1 for highly diluted solutions.

The same happens at the second electrode, but due to a different material and consequently a different chemical reaction, the electric potential difference will not be the same. The difference in the two potential jumps is the equilibrium cell voltage (also called open-circuit voltage, *OCV*), which can be measured at the two electrodes. For many sorts of batteries comprising metal electrodes, Eq. (4.7) is the cause for a voltage decrease during discharge. At the anode, c_{OX} continuously rises, and E_{anode} therefore increases. At the cathode, c_{OX} is used up, and $E_{cathode}$ decreases. The total cell voltage $V = E_{cathode} - E_{anode}$ decreases. If this decrease is sufficiently high, the open circuit voltage can be used to determine the state of charge of the battery.

Once an external electrical connection is set up, electrons start to move from the more negative electrode to the positive one (see Fig. 4.5). The lack of electrons on the negative electrode (the anode) causes the chemical reaction to consume the reduced species in order to regain equilibrium, while the opposite occurs at the positive electrode (the cathode). The obtained current I depends on the potential displacement η (in V, also called overpotential), as well as the reaction rate, usually expressed by the exchange current density i_0 (in Am^{-2}). For electrode x, this relationship is defined by the *Butler-Volmer Equation*:

$$I_x = A_x i_0 \left[\exp \left(\frac{\alpha_{a,x} n_x F \eta_x}{RT} \right) - \exp \left(-\frac{\alpha_{c,x} n_x F \eta_x}{RT} \right) \right], \tag{4.8}$$

with A_x ...active inner electrode surface (in m^2), α_a and α_c ...symmetry coefficients of the chemical reaction in anodic (left-to-right) and cathodic direction (right-to-left) respectively. These are derived from Transition-State Theory and often set to 0.5, which indicates perfect symmetry. The total current is now given on the one side by the external load, in the case of a resistor this would follow *Ohm's Law* ($I = V/R$), and on the other side by the reactions on the electrodes x and y, whereby

$$I = i_x = i_y, \tag{4.9}$$

and

$$V = E_x - E_y - (\eta_x - \eta_y). \tag{4.10}$$

4.4.2 Ion Transport

Normally, the charge transport in the electrolyte is done by a single ion species solely, e.g. Li^+ in lithium ion or OH^- in nickel metal-hydride cells. The counter ions, e.g. PF_6^- and K^+ respectively in the cell types mentioned, do not participate in the reactions, however supply neutrality. The active ions are produced on one electrode and consumed on the other, causing a concentration gradient in the electrolyte, whereby the counter ions follow the active ones. The transport mechanisms of diffusion, migration and convection reduce this gradient continuously. Convection can be found in systems with larger electrolyte compartments such as some lead-acid systems or fuel cells with liquid electrolyte, but open spaces in batteries are usually too small to provide good convection. Migration is the transport of ions in the electric field between the electrodes. As the ions are usually surrounded by solvent molecules shielding them, the effect is not very strong. However, close to solid spikes at high overpotentials (e.g. during charging at high rates), this field can get very high and cause ions to be reduced especially at these spikes, causing the development of dendrites.

The main transport mechanism is diffusion. Neglecting a possible hindrance of slow-moving counter ions, the *Nernst-Planck Equation* describes the rate of diffusion:

$$\frac{\partial a_i}{\partial t} = \nabla \cdot \left(D_i \nabla a_i + \frac{zF}{RT} D_i a_i \eta \right), \tag{4.11}$$

with $a_i = \gamma_i c_i$ ...activities of species i, D_i ...diffusion coefficient.

4.4.3 Electron Transport

In the electrodes and the current collector, the charge is commonly transported by electrons. For the electronic current, *Ohm's Law* describes the process:

$$-\nabla \cdot (\sigma \nabla \phi) = 0, \tag{4.12}$$

with σ ...electrical conductivity and ϕ ...electrical potential. Although the current collectors usually consist of solid metals, either low conductivity (e.g. lead grid in lead-acid batteries) or low thickness (e.g. thin Cu foils of $10\,\mu m$ in lithium-ion batteries) can significantly increase the total internal resistance.

4.4.4 Porous Electrodes

In order to increase the inner surface A_i, battery manufacturers use highly porous active layers in their cells, as schematically shown in Fig. 4.5. While this has a positive impact on the charge transfer rate, it is a hindrance for diffusion: The smaller the particles, the better the current exchange, but the worse the mass transport. A_i is a parameter already existing in Eq. (4.8), the ion transport Eq. (4.11) contains the bulk variable D_i. Already in 1935, *Bruggemann* proposed an empirical way to calculate an effective value $D_{i,eff}$:

$$D_{i,eff} = D_i \varepsilon_i^{brug_i}, \tag{4.13}$$

with i ...material of interest (anode, cathode, separator), ε_i ...porosity and $brug_i$ ...*Bruggemann Coefficient*, which is commonly set to 1.5 for spherical particles. In other words, ε_i represents the electrolyte ratio, while $brug_i$ represents the degree of tortuosity.

In addition, the same approach can be used for the electric conductivities in solid (σ_i) and liquid (κ_i), whereby ε_i in the first case represents the solid portion of the porous material:

$$\sigma_{i,eff} = \sigma_i \varepsilon_i^{brug_i}, \tag{4.14}$$

$$\kappa_{i,eff} = \kappa_i \varepsilon_i^{brug_i}. \tag{4.15}$$

Another approach could be the development of a 3D-structure and calculation of the flow characteristics, which is much more complex however. The big advantage, on the other hand, would be the ability to calculate any form of particles.

4.4.5 Intercalation

In addition to the aforementioned aspects, the lithium-ion technology includes a special feature: ion intercalation. In contrast to most other cell types, the active ion (Li^+) is not deposited on the electrode surface, but stored inside a host material. For anodes, graphite is usually chosen, whereas for cathodes a number of metal oxides and phosphates are available. Section 4.4.1 described how the ion concentration influences the cell voltage and consequently the degree of cell discharge (state of charge), whereby it was assumed, that the electrode does not change. In the case of host materials, this does not hold true. Regarding the mechanism of ion uptake (and release), we can distinguish between two types of host materials: energy and entropy-controlled systems.

In energy-controlled systems (see Fig. 4.6), such as graphite or many oxides, all particles fill up in parallel, and a "concentration" of active ions develops in the host material (c_s). For example graphite (C_6) can take up $1/6$ th molar ratio of lithium (LiC_6) at most, and the lithium content is usually between 0 and 1, written as Li_xC_6, $0 < x < 1$, continuously. For graphite, c_s and x are connected linearly:

$$c_s = \frac{x \cdot \rho_C}{6 \cdot M_C} \approx 30x, \tag{4.16}$$

Fig. 4.6 Schematic view
of particles at different
state of charge; *top row*
energy-controlled, *bottom
row* entropy-controlled

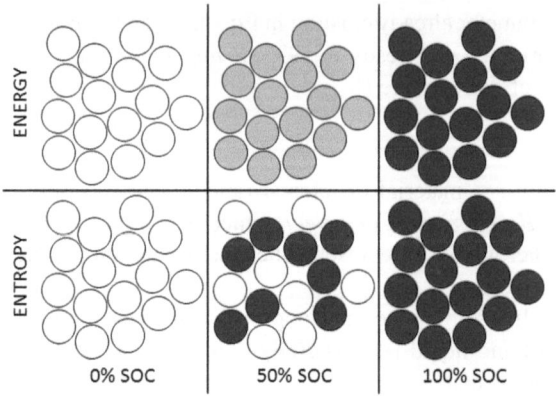

with c_s in molL^{-1}, ρ_C ...density of graphite ($\approx 2.2 \cdot 10^3$ gL^{-1}) and M_C ...molar
weight of graphite (≈ 12.01 gmol^{-1}). Hence, the maximum concentration in the
particle is $c_{s,max} \approx 30$ molL^{-1}. The solid diffusion into the particles can also be
addressed using a diffusion coefficient D_s and a radial coordinate r (spherical particles assumed):

$$\frac{\partial c_s}{\partial t} = \frac{1}{r^2} \frac{\partial}{\partial r} \left(D_s r^2 \frac{\partial c_s}{\partial r} \right). \tag{4.17}$$

In lithium-ion batteries, the total concentration of Li^+ in the electrolyte (c_L) remains
the same (equilibrium condition); hence, the state of charge and the related potentials
E_i depend solely on the concentrations of the lithium in the solids.

In entropy-controlled systems (Fig. 4.6, bottom row), the particles fill up sequen-
tially, resulting in a mixed potential at the current collector. These systems often
show flat discharge curves (voltage versus state of charge or versus average lithium
concentration). One example is lithium-iron phosphate. There are simple modelling
approaches that describe the dependence of the open circuit voltage simply by the
ratio of filled to empty particles, provided the particles are small and the inner-particle
diffusion rate is negligibly small. More sophisticated approaches use e.g. the *Cahn-
Hilliard equation* or address the crystallographic characteristics of the host material,
thereby supporting diffusion of lithium in one, two or three directions (1D, 2D or 3D
materials, respectively [10]).

4.4.6 Heat Generation

Heat is a by-product of cell operation that is mostly unwanted (energy loss). Temperature increase of the battery can lead to faster ageing and to safety-critical battery
states and must be avoided by implementing cooling strategies. Heat generation is
commonly modelled by the energy balance comprising external heating or cooling,

heating caused by current flowing and heating or cooling by chemical reactions. The heat equation is generally written as

$$\rho c_p \frac{\partial T}{\partial t} = \nabla \cdot (\lambda \nabla T) + f, \tag{4.18}$$

with ρ ...density, c_p ...heat capacity, λ ...thermal conductivity, and f ...sum of heat sources. Here heat transport is assumed to be caused only by conduction and it is covered by the first term on the right (*Fourier's Law*). Heat caused by the current (*Joule heating*) is expressed as:

$$f_{Joule} = \sigma \nabla \phi \cdot \nabla \phi. \tag{4.19}$$

The heat caused by chemical reactions can also be calculated by a similar relation:

$$f_{react} = I\eta, \tag{4.20}$$

with I ...exchange current(see Eq. (4.8)) and η ...overpotential. The very simple approaches shown here can be modified by using more complex descriptions to address open circuit voltage drifts, conductivity changes etc. with temperature (e.g. *Arrhenius*) or effects like *Seebeck/Peltier* or *Soret*.

4.4.7 Cell Ageing

Degradation modelling is another important field. Usually, the decay of battery capacity and the increase of the internal resistance are of interest. Empiric or semi-empiric approaches are used, similar to the models in Sect. 4.2, which now describe the ageing effects. Also, attempts to cover multiple influencing factors on larger experimental scale and setting up a statistical model were done (e.g. [9]), still being empirical. Mechanistic ageing models are physics-based (Sects. 4.4.1 to 4.4.5), whereby time-dependent parameter modifications are added. However, in many cases the ageing mechanisms are not available together with a selective quantification of the underlying parameters.

4.5 Large-Scale Modelling

At a certain point in the modelling process, a 1D cut through a cell stack might no longer be sufficient. Especially for optimization of cell, module and pack design, there is high need for 3D simulation. In 3D simulation, the most common methods all depend on computational grids that map the 3D structure of the simulation domain to a finite number of nodes (coordinate points), making the simulated object accessible for

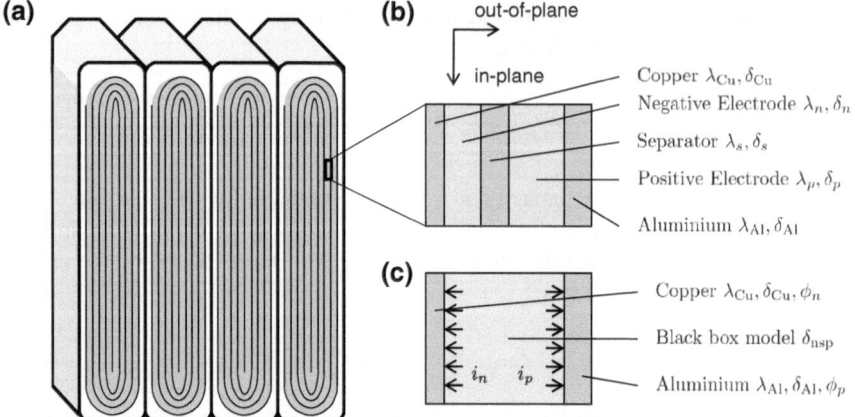

Fig. 4.7 a Schematic structure of battery module. **b** Structure of stack layer. **c** *Black-box* approach to electric behaviour

computer algorithms. For example, in the finite element method, a three-dimensional domain is resolved by simple geometrical bodies such as tetrahedrons, pyramids and hexahedra. The computational cost is thereby increasing with every node added to the grid.

One constraint on the placement of these elements is that their surfaces have to map the boundaries and interfaces of different materials and jumps of physical properties in general. For example, industrial lithium-ion battery modules consist of several cells packed together (see Fig. 4.7a), which are structured in layers (see Fig. 4.7b), either stacked on each other, wound prismatically or cylindrically, forming the "jelly roll". Resolving this structure layer by layer makes it possible to assign material properties to the current collectors and the cell stack separately. However, this would require quite an expensive computational effort due to the sheer number of stacked layers.

In this situation, a homogenization of the properties of the jelly roll is preferred. Homogenization means scaling up equations such that the small-scale changes of physical quantities can be averaged and allow to resolve the simulation domains on a coarser computational grid. An approach such as this allows the jelly roll to be geometrically resolved independent from the number of layers and their geometrical borders. The material properties assigned to the domain represent the averaged or homogenized properties of all the components of the jelly roll.

4.5.1 Thermal Behaviour

A simple example can be given with the homogenization of the heat equation in the layered cell.

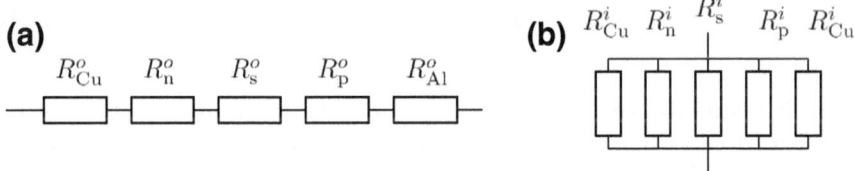

Fig. 4.8 Thermal RC equivalent circuit of electrode stack

For the following procedure a periodic structure, as indicated in Fig. 4.7b, is assumed. A rigorous mathematical derivation of the homogenization can be found, for example, in [4]. Here, only an intuitive argumentation will be given.

The aim now is to homogenize the heat Eq. (4.18) and derive a homogenized equation of the form

$$\widehat{c_p \rho} \hat{T}_t - \nabla \hat{\lambda} \nabla \hat{T} = \hat{f}, \quad \text{for } x \in \Omega, \tag{4.21}$$

where $\Omega \subset \mathbb{R}^3$ is the 3D domain of the jelly roll and the $\hat{\ }$ operator describes homogenized quantities, which should describe averaged values of the stack composite, instead of resolving every material jump separately. Often, these quantities are called effective values in this context. For the middle part of the jelly roll, which comprises only parallel layers, a spatially constant $\hat{\lambda}$ can be expected because the stacked structure does not change there. This does not hold true for the curved parts, where the definition of in-plane and out of-plane is dependent on the position in the jelly roll. Obviously, the average conductivity of the jelly roll will be different in the in-plane and the out-of-plane directions, as indicated in Fig. 4.7b. The thermal resistances in the out-of-plane direction can be compared to an electrical circuit of resistors in series (see Fig. 4.8a) because a possible heat flow would have to go through the layers consecutively. Here, the area-specific conductivity G^o (in Sm) of a single layer depends on the specific conductivity λ (in Sm^2) of the material and its thickness δ by

$$G^o_{\text{layer}} = \frac{\lambda_{\text{layer}}}{\delta_{\text{layer}}}. \tag{4.22}$$

The total conductivity of resistors in series is calculated by the harmonic mean of the single conductivities, so

$$G_{\text{stack}} = \left(\sum_{textlayer} \frac{1}{G_{textlayer}} \right)^{-1}. \tag{4.23}$$

With L being the total stack thickness, one can now derive the specific conductivity as

$$\lambda_{\text{stack}}^o = L \cdot G = L(\sum_{layer} \frac{\delta_{\text{layer}}}{\lambda_{\text{layer}}})^{-1}, \tag{4.24}$$

which is exactly the weighted harmonic mean of the conductivities. The in-plane direction can analogously be compared to a set of resistors in parallel (see Fig. 4.8b), because a heat flow would go through the layers in parallel. Based on the equivalent circuit, one can once again easily derive the specific stack conductivity as the weighted arithmetic mean of the single layer conductivities:

$$\lambda_{\text{stack}}^i = \frac{1}{L} \sum_{layer} \delta_{\text{layer}} \lambda_{\text{layer}}. \tag{4.25}$$

Notice that the in-plane conductivity of a layer grows with its thickness, as opposed to the out-of-plane case, where it shrinks with it. All together, this anisotropic behaviour is described by the diffusion tensor

$$\hat{\lambda} = \begin{bmatrix} \lambda_{\text{stack}}^o & 0 & 0 \\ 0 & \lambda_{\text{stack}}^i & 0 \\ 0 & 0 & \lambda_{\text{stack}}^i \end{bmatrix}, \tag{4.26}$$

assuming that the cell is positioned with its out-of-plane direction in the x-axis. Remarkably, with these two cases of a layered structure, the two extreme cases of effective homogenized conductivities have been addressed. Theory shows that any arbitrary geometrical composition that has the same volumetric distribution of materials yields effective conductivities that lie between the arithmetic and harmonic weighted averages of their components.

Independent of the geometrical structure of the representative stack, the effective capacity $\widehat{c_p \rho}$ is the arithmetic mean of its components weighted by their volume fraction:

$$\widehat{c_p \rho} = \sum_i V_i c_{pi} \rho_i, \tag{4.27}$$

with i ...the component of the stack, V_i ...its volume fraction.

4.5.2 Electrical Behaviour

For the homogenization of the electrical behaviour of the cell, an assumption about the distribution of anodic and cathodic currents is proposed, which is that the anodic and cathodic current densities to the current collectors in any point of the jelly roll are anti-symmetric to each other. In this context, anti-symmetric means that the cathodic current at any point of the positive current collector is the negative of the anodic current at the opposite point on the negative current collector. This assumption is supported by the high collector foil conductivities compared to the lower electrode conductivities.

Under this assumption, the electric potentials in the current collector foils can be modelled with the following set of equations:

$$\begin{aligned}
-\nabla \cdot \sigma_{Cu} \nabla \phi_n &= 0, & \text{for} \ \ x \in \Omega_{Cu}, \\
-\nabla \cdot \sigma_{Al} \nabla \phi_p &= 0, & \text{for} \ \ x \in \Omega_{Al}, \\
\sigma_{Cu} \nabla \phi_n &= i, & \text{for} \ \ x \in \Gamma_{Cu}, \\
\sigma_{Al} \nabla \phi_p &= -i, & \text{for} \ \ x \in \Gamma_{Al},
\end{aligned} \tag{4.28}$$

where i ...current density at the collector-electrode interfaces Γ_{Cu} and Γ_{Al}, σ ...electric conductivity of the respective material, ϕ_n and ϕ_p ...pair of electric potentials.

Leaving the behaviour between the two collector foils to a black-box model, as indicated in Fig. 4.7c, and following the procedure from above, a set of equations in the form

$$\begin{aligned}
-\nabla \cdot \hat{\sigma}_{Cu} \nabla \hat{\phi}_n &= \hat{i} \ \ \text{for} \ \ x \in \Omega, \\
-\nabla \cdot \hat{\sigma}_{Al} \nabla \hat{\phi}_p &= -\hat{i} \ \ \text{for} \ \ x \in \Omega,
\end{aligned} \tag{4.29}$$

can be derived.

Through this step, the computational domain is highly simplified because the collector foils no longer have to be distinguished in a geometrical way. This simplification yields a big reduction in computational cost because the spatial discretization no longer needs to resolve the individual layers of the jelly roll, even though the positive and negative potentials are distinguished by the potential pair $(\hat{\phi}_p, \hat{\phi}_n)$. Thus, the set of Eq. (4.29) represents a continuous domain, where at every point the positive and the negative potential exist and represent the potential pair at a typical electrode stack at this macroscopic point. The calculation of the effective conductivity follows the one for the heat equation, with the difference that the electrical insulation between the two foils yields zero conductivity normal to the stack layers. Therefore in the parallel part of the jelly roll, the conductivity is

$$\hat{\sigma}_{Cu} = \begin{bmatrix} 0 & 0 & 0 \\ 0 & \sigma_{Cu}^i & 0 \\ 0 & 0 & \sigma_{Cu}^i \end{bmatrix}, \hat{\sigma}_{Al} = \begin{bmatrix} 0 & 0 & 0 \\ 0 & \sigma_{Al}^i & 0 \\ 0 & 0 & \sigma_{Al}^i \end{bmatrix}. \tag{4.30}$$

4.5.3 Distributed-Micro-Structure Modelling

An open question is how the source terms $\hat{f}$ (heat production) and $\hat{i}$(current density) in Eqs. (4.21) and (4.29) are calculated in the model. Of course, this depends mainly on the black-box model chosen for the electrochemical part of the stack. However, the question is how this micro model is linked to the macroscopic model. The ideas of the heterogeneous multi-scaling method (HMM) [1] and a work by Kim et al. [6] form the basis for the following ideas.

For the current density, the relation between the microscopic current density i and the macroscopic current density $\hat{i}$ is simply given by the relation

$$\hat{i} = A_e i \tag{4.31}$$

with A_e (in $\frac{m^2}{m^3}$) being the collector foil surface-to-cell-volume ratio. This means that the collected current per area (in Am^{-2}) is scaled by the collector foil area to a current source per volume (in Am^{-3}). Since the contribution of the black-box model to the heat source $\hat{f}$ should already be calculated per unit volume, there is no need for scaling. Equations (4.21) and (4.29) show that at every point x in the jelly roll the information on the source terms i and f is needed to calculate the macroscopic potential and temperature distribution. Only the simplest models would allow for a formulation of the dependence of the source terms on the potential pair $(\hat{\phi}_p, \hat{\phi}_n)$ and the temperature T, which could be stated in Eqs. (4.21) and (4.29) explicitly.

Following the ideas of HMM, a distributed-micro-structure model can be set up, instead of oversimplifying the black-box model. This means that at every point where the information on the source terms is needed (the macroscopic nodes, see Sect. 4.5), a micro simulation of the black-box model is set up, which calculates the source terms for the macro model in dependence on the macroscopic variables.

Furthermore, the points where the micro structure is simulated can be decoupled from the points of the macroscopic numerical grid by interpolating the information between an arbitrary chosen set of points of simulated micro structure and the set of points of the numerical grid.

In this way a hierarchical order of models and sub-models is created that allows for the coupling of the micro structure of the electrode with the macro structure of the jelly roll. This reduces the computational effort, and enables the simulation of whole battery modules and packs with an adjustable degree of detail.

Acknowledgments The authors would like to acknowledge the financial support of the "COMET K2 - Competence Centres for Excellent Technologies Programme" of the Austrian Federal Ministry for Transport, Innovation and Technology (BMVIT), the Austrian Federal Ministry of Economy, Family and Youth (BMWFJ), the Austrian Research Promotion Agency (FFG), the Province of Styria and the Styrian Business Promotion Agency (SFG).

The research leading to these results has received funding from the European Community's Seventh Framework Programme (FP7/2007-2013) under Grant Agreement no 266090 (SOMABAT).

References

1. Abdulle A, Weinan E, Engquist B, Vanden-Eijnden E (2012) The heterogeneous multiscale method. Acta Numer 21:1–87. doi:10.1017/S0962492912000025
2. Bard AJ, Faulkner LR (2001) Electrochemical methods - fundamentals and applications, 2nd edn. Wiley, New York
3. Hamann C, Vielstich W (2005) Elektrochemie. Wiley, New York. http://books.google.at/books?id=TLXeAAAACAAJ

4. Hornung U (1997) Homogenization and porous media. Interdisciplinary applied mathematics. Government Printing Office, Washington, U.S
5. Kauffman GB (2009) Electrochemical impedance spectroscopy (Mark E. Orazem and Bernard Tribollet). Angewandte Chemie Int Ed 48(9):1532–1533. doi:10.1002/anie.200805564
6. Kim GH, Smith K, Lee KJ, Santhanagopalan S, Pesaran A (2011) Multi-domain modeling of lithium-ion batteries encompassing multi-physics in varied length scales. J Electrochem Soc 158(8):A955–A969. doi:10.1149/1.3597614
7. Linden D, Reddy T (2002) Handbook of batteries, 3rd edn. McGraw-Hill, New York
8. Newman J, Thomas-Alyea KE (2004) Electrochemical systems, 3rd edn. Wiley, Hoboken. ISBN:978-0471477563
9. Prochazka W, Pregartner G, Cifrain M (2013) Design-of-experiment and statistical modeling of a large scale aging experiment for two popular lithium ion cell chemistries. J Electrochem Soc 160(8):A1039–A1051. doi:10.1149/2.003308jes, http://jes.ecsdl.org/content/160/8/A1039.full.pdf+html
10. Safari M, Delacourt C (2011) Mathematical modeling of lithium iron phosphate electrode: Galvanostatic charge/discharge and path dependence. J Electrochem Soc 158(2):A63–A73. doi:10.1149/1.3515902
11. The Engineering ToolBox (2013) Data downloaded at the 26.09.2013. Online, http://www.engineeringtoolbox.com/lead-acid-battery-d_1544.html

Chapter 5
Analytical Methods for Investigation of Lithium-Ion Battery Ageing

Sascha Weber, Sascha Nowak and Falko Schappacher

Abstract One of the major issues battery research must address is the lifetime of a cell. This can be reduced by physical and chemical ageing processes that occur inside the cell and are influenced by both the operating strategy and the surrounding conditions (e.g. temperature). To understand battery ageing, it is necessary to analyze the materials used in a cell at the microscopic level and correlate the results with electrical measurement data. This chapter describes a strategy for performing an ageing experiment by using a combination of analytical methods.

Keywords Lithium-ion battery · Ageing · Analytical methods

5.1 Introduction

Today, lithium-ion batteries (LiBs) are recognized as the state-of-the-art technology for portable energy storage (e.g. for mobile phones, cameras or notebook PCs). The LiB could play an even more important role by helping to introduce electro mobility for the masses, due to their high gravimetric and volumetric energy density as well as – depending on the choice of materials – the potential to deliver high power and provide high safety.

S. Weber (✉) · S. Nowak · F. Schappacher
Westfälische Wilhelms-Universität Münster, MEET Battery Research Center,
Corrensstr. 46, 48149 Münster, Germany
e-mail: sascha.weber@uni-muenster.de

S. Nowak
e-mail: sascha.nowak@uni-muenster.de

F. Schappacher
e-mail: falko.schappacher@uni-muenster.de

A. Thaler and D. Watzenig (eds.), *Automotive Battery Technology*,
Automotive Engineering: Simulation and Validation Methods,
DOI: 10.1007/978-3-319-02523-0_5, © The Author(s) 2014

Fig. 5.1 Schematic drawing of the working principle of a lithium-ion battery. To reduce complexity, the SEI is not shown here

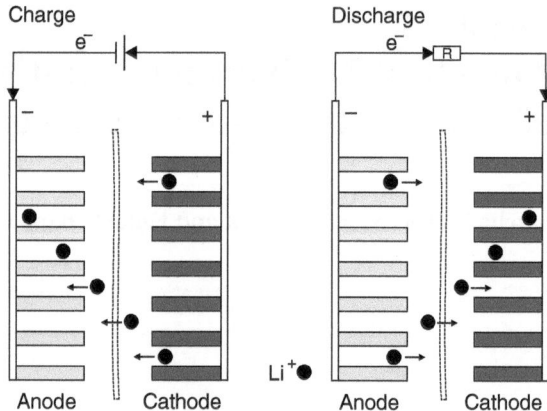

5.1.1 Working Principle of a Lithium-Ion Battery

A typical LiB consists of two electrodes (anode and cathode), which are composite materials coated on a metal foil (the current collector, copper or aluminum), and a separator that is soaked in an organic electrolyte, all of which is surrounded by an airtight housing. The composite electrodes commonly contain the active material that can serve as a host compound for lithium ions, a conductive agent to optimize electronic conductivity, and a binder that holds everything together. The most common active material used on the anode side is graphite, while transition metal oxides such as $LiCoO_2$ and many others can be used on the cathode side.

The working principle of a LiB is based on a so-called 'rocking chair' mechanism, which is illustrated in Fig. 5.1. During charge, lithium ions migrate from the cathode side to the anode side of the cell when an external current is applied. During discharge, the lithium ions migrate from the anode to the cathode side, and the cell delivers a current. The ionic conductivity is provided by the organic electrolyte, which includes a conductive lithium salt, e.g. $LiPF_6$. The electrode/electrolyte interfaces are not inert, but rather feature a versatile surface chemistry. Especially on the anode side, the organic electrolyte will undergo decomposition reactions leading to the formation of a surface film that consists of organic and inorganic compounds, which is called the solid-electrolyte interface (SEI). For more details on this basic principle please refer to the following introductory articles [30, 31].

5.1.2 Ageing of Lithium-Ion Batteries

One of the major issues of current LiB technology is the loss of capacity over time which is known as ageing. In theory, two types of ageing have to be differentiated. Calendar, or storage life, describes ageing effects that occur during the rest phases

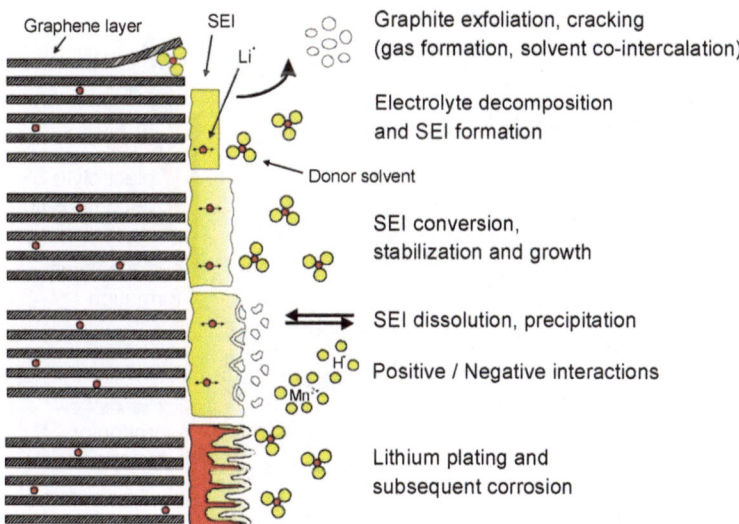

Fig. 5.2 Ageing processes occurring at the anode/electrolyte interface (Reprinted from [26] (Fig. 1), Copyright 2005, with permission from Elsevier)

of a battery (when it is stored in a certain state), whereas cycle life stands for all ageing effects that occur when current is applied (i.e. the battery is in use). In reality, both cycle life and calendar life are mixed and result in a complex ageing behavior, depending on the mode of use of the battery.

Due to the large variety of components used in a lithium-ion battery, it is impossible to give a complete overview of all ageing mechanisms that are likely to occur in a specific cell setup. However, some general hints can be extracted from the relevant literature. Figure 5.2 provides a general overview of ageing mechanisms related to the anode. The solid-electrolyte interface (SEI) is formed initially by the decomposition of electrolyte components during cell formation. Subsequently, both SEI conversion and SEI dissolution can occur. Furthermore, the layered structure of graphite can be destroyed by exfoliation.

On the cathode side, ageing is more related to structural changes of the active material and degradation of the electrode indicated as loss of contact, micro-cracking and oxidation of conductive particles, as shown in Fig. 5.3.

A review published by Barré et al. [2] ranks the three main ageing mechanisms known to literature: (1) loss of active lithium due to SEI forming and converting reactions, as well as other side reactions; (2) loss of active electrode material due to material dissolution, particle isolation and structural degradation; and (3) resistance increase in the cell due to contact loss inside the electrodes, as well as electrolyte degradation.

To understand LiB ageing on a microscopic level that is beyond the measurement of an electronic resistance, it is necessary to develop and apply analytical methods

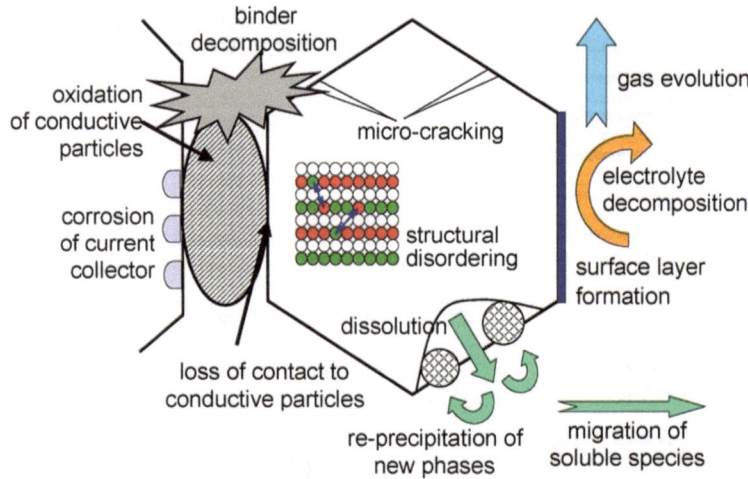

Fig. 5.3 Overview of basic ageing mechanisms of cathode materials (Reprinted from [26] (Fig. 2), Copyright 2005, with permission from Elsevier)

that are able to measure the ageing effect under investigation. Furthermore, methods have to be developed to extract the materials of interest from the cell without contamination. This chapter offers an insight into the possibilities of some analytical methods that are a type of standard for ageing investigations. However, every method has its limitations, and it must be adjusted to meet the requirements of a specific sample.

5.1.3 Investigating Cells

Generally, ageing investigations are based on the comparison of chemical and physical characteristics of fresh and aged material. Depending on the availability of the material used in the cells under investigation, this can involve raw materials such as powders used for making the electrodes, fresh electrodes, cells after formation by the manufacturer and aged cells. To obtain statistically representative data, the load profile, including other influencing factors under investigation (e.g. temperature), should be developed using the design-of-experiment method (DoE), as described by Prochazka et al. [17].

Figure 5.4 shows a general flow chart, including the possibility of obtaining raw materials from the cell manufacturer. The first two columns represent the initial state, taking into account that, in most cases, raw materials will not be supplied by a cell manufacturer. The third column represents the aged state which is achieved by applying a load profile to the cells.

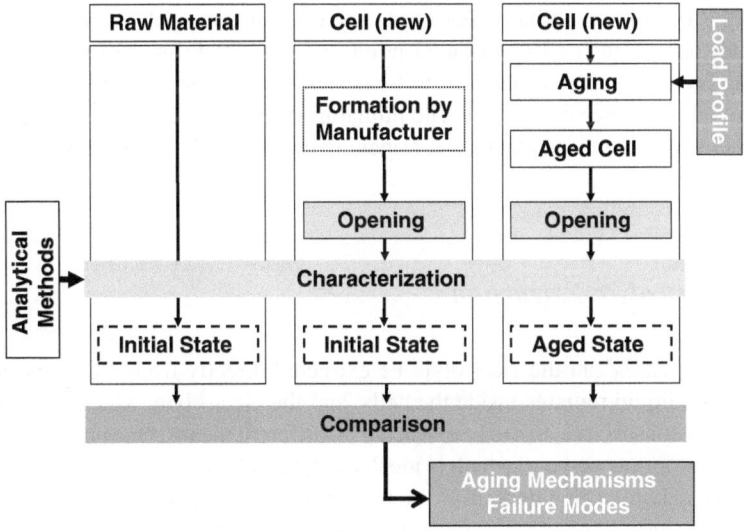

Fig. 5.4 Flow chart of a typical ageing experiment

5.2 Harvesting Material from Cells

The workflow for extracting the material to be analyzed from the cell can be divided into several steps. First, the cell has to be brought to a safe state, which is normally defined as completely discharged using procedures provided by the cell manufacturer.

5.2.1 Cell Opening

Opening of LiB cells is highly dangerous and should only be executed by trained people! Small mistakes may lead to fire or even explosion. During cell opening, it is useful to have a Dewar vessel with liquid nitrogen within reach to extinguish small fires and get the heat out of the system.

Before opening any cell, it is important to know about the internal design of the cell to prevent short circuits and hazards during cell opening. X-ray photographs of the cell provide very useful information about the internal cell set-up. In general, three different cell geometries and housings are used in the field of LiBs: cylindrical or prismatic hard-case cells and prismatic aluminum-laminate pouch-bag cells. Since LiBs consist of materials that are highly sensitive to moisture or air (e.g. the widely used conducting salt $LiPF_6$), it is essential to develop cell opening procedures that can be carried out under inert gas conditions.

Pouch-bag cells can easily be opened in a glove box using a ceramic knife and ceramic scissors. In this case, contamination of the cell materials caused by the opening procedure is unlikely. For hard-case cells, however, the contamination of the inner part of the cell is a crucial point, due to dust from abrasion. Thus cell opening procedures and tools have to be developed that can be applied in a glove box and only produce coarse grit. The miniaturization of tools found in mechanical workshops (e.g. a turning machine for opening cylindrical hard-case cells) provides good results.

5.2.2 Electrolyte Extraction

Only in rare cases can the electrolyte be extracted directly from a cell. Normally, there are no liquid remains inside the cells, and the electrolyte, which sticks to the cell component surfaces, has to be extracted by different means. One option is rinsing with an appropriate solvent, which is ideally not part of the electrolyte composition. The disadvantage of this method is the high dilution factor, which results in levels of decomposition products and/or additives that are below the limit of detection for the planned analytical methods. Furthermore, solid parts of the cell will be dissolved and can hinder the subsequent measurements.

A more recent approach for the extraction of the electrolyte from cells is the usage of supercritical CO_2. This method is faster, more selective and efficient than extraction with solvent [20, 22]. Moreover, no sample pre-concentration or clean-up process is necessary [6]. Additionally, through-flow experiments can enhance the recovery rate, but the concentration of the conducting salt in the extract depends on the adsorption properties of the extracted material. Consequently, the usage of supercritical CO_2 is a suitable method for the separation of the electrolyte from other battery parts.

5.2.3 Sampling of Electrodes

Depending on the cell design, different approaches for the sampling of parts of electrodes can be chosen. In the case of common lab-style cells (e.g. coin cells or small pouch cells), the electrode area is too small to extract spatially resolved samples. When dealing with larger cells (e.g. wound or stacked types) it could be of interest to extract samples at the beginning, middle and end of a wound electrode.

Depending on the analytical investigation, a washing step may be necessary. This is the case, for example, if the sample needs to be handled in air, and the decomposition of residues of the conductive salt would damage the sample.

Experiments performed using X-ray photoelectron spectroscopy (see Sect. 5.3.1) show the influence of different washing steps on the SEI. Commercial graphite anodes were held in an argon atmosphere and then immersed in dimethyl carbonate (DMC) for one second or one minute. Comparison of the unwashed sample and the two washed samples (see Fig. 5.5) suggests that the organic part (e.g. $R-CH_2OCO_2Li$)

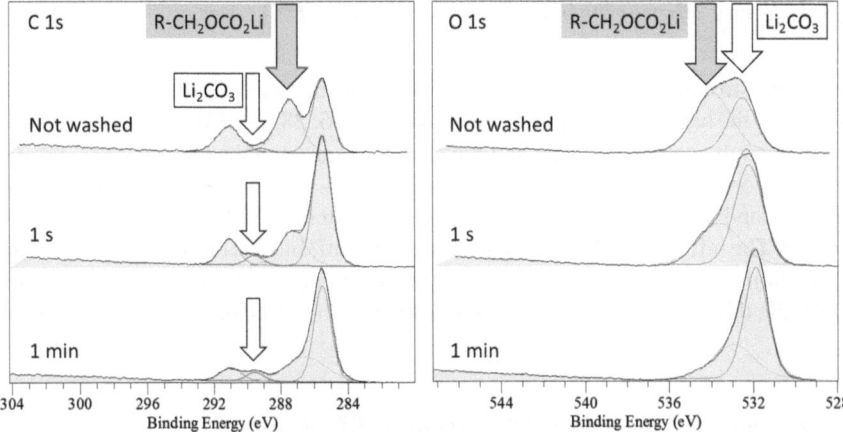

Fig. 5.5 XPS spectra showing C 1 s and O 1 s core scans of the SEI on graphite anode material, prepared without washing, washed in dimethyl carbonate (DMC) for 1 s and for 1 min. Variations of the signals belonging to the inorganic Li_2CO_3 component and to the organic $R–CH_2OCO_2Li$ component are indicated as *white* and *gray* arrows, respectively

of the SEI should be removed, whereas inorganic Li_2CO_3 remains on the surface. A detailed investigation using sputter-depth profiling was conducted by Niehoff et al. [13], which showed that detailed information about the thickness and structure of the anode SEI can be obtained without washing the sample.

For this method, conservation of the SEI was made possible by omitting a washing step as air contact of the sample could be avoided during sample handling. In the case of other analytical methods, a careful pre-investigation could proved further information regarding the necessity of a washing step during sample processing.

5.3 Analysis of Electrodes

There are two main important issues when dealing with the ageing of the electrodes: impurities and various reactions and processes that occure during cycling. The battery materials have to be free of impurities because residues of the production process can have various negative impacts on the performance of LiBs [35], such as reduced cycling stability or loss of energy density. Impurities or metal ions may also be dissolved during the charge/discharge process [34], migrate into the electrolyte and then damage either the SEI or one of the electrodes. Another source of performance decreasing ions/impurities is the current collector, e.g. corrosion of aluminum [36].

5.3.1 X-ray Photoelectron Spectroscopy (XPS)

LiBs rely strongly on the SEI. The SEI forms at the anode and cathode, thereby protecting the battery from further degradation [3, 16, 28, 29, 33]. The SEI formation depends on the active materials and the electrolyte used in the battery [7, 9, 18, 19, 27, 32]. For analyzing the ageing mechanisms of LiBs, the characterization of the structure and composition of the SEI is of great interest. XPS is a surface-sensitive method. The information depth is in the range of 3 nm, and it is one of the few methods where lithium can be directly analyzed. Using sputter-depth profiling XPS (SDP-XPS), the thickness, structure and composition of the SEI can be investigated [13].

5.3.2 Scanning Electron Microscopy (SEM) and Energy-Dispersive X-ray Spectroscopy (EDX)

SEM is an image-based method and a very useful tool for investigating the ageing effects of LiBs. Low magnifications give an overview of the electrode surface. Cracking of the electrode due to binder disintegration can easily be observed. At higher magnifications, the electrode materials can be investigated. To visualize the SEI or the binder, it is important to choose a proper accelerating voltage. At high accelerating voltages, organic structures such as the SEI or the binder might not be seen or can even be destroyed by the electron beam. Figure 5.6 shows images of an aged graphite-based anode at magnifications of 50,000 times (inset) and 100,000 times. The inset in Fig. 5.6 shows the layered structure of the graphite. The spherical particles with a diameter of approximately 40 nm are typical for a conductive agent such as *Super P®Li*. The image taken at a magnification of 100,000 times shows the web-like structure of the PVDF-based binder and the SEI that grew during cycling.

Energy-dispersive X-ray spectroscopy (EDX) analyses the X-rays emitted during SEM for the elemental analysis or chemical characterization of the specimen [5]. With EDX, it is possible to analyze striking particles or areas, or even to map the whole section, as shown in Fig. 5.7. The EDX mapping reveals that the cathode consists of two different active materials. In addition, EDX of the individual particles shows the composition of the different active materials: $LiNi_{0.5}Mn_{0.3}Co_{0.2}O_2$, and $LiCoO_2$.

5.3.3 Elemental Analysis (ICP, TXRF)

The distribution of the specific impurities or reaction processes on the surface of the electrodes (and especially the quantification in the bulk material) can be characterized using Inductively Coupled Plasma (ICP) techniques or techniques such as total reflection X-ray fluorescence (TXRF).

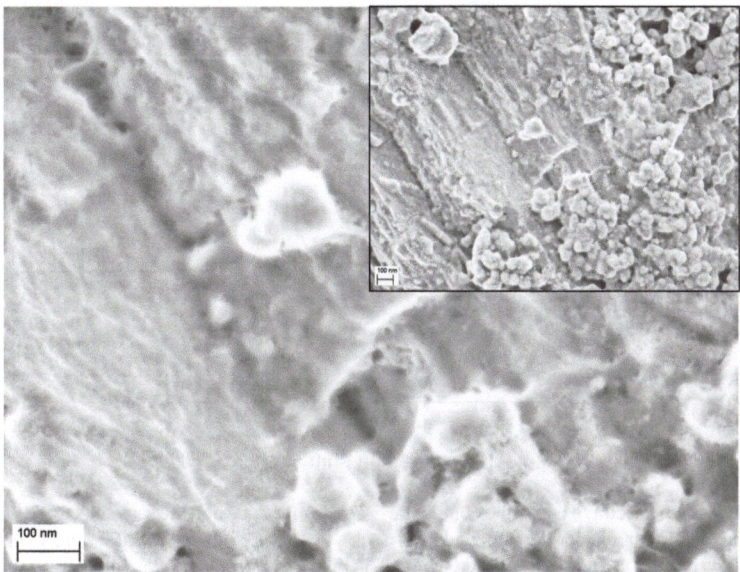

Fig. 5.6 SEM image of a graphite-based anode taken at 1 kV and a magnification of 100,000 times. The *inset* shows a magnification of 50,000 times

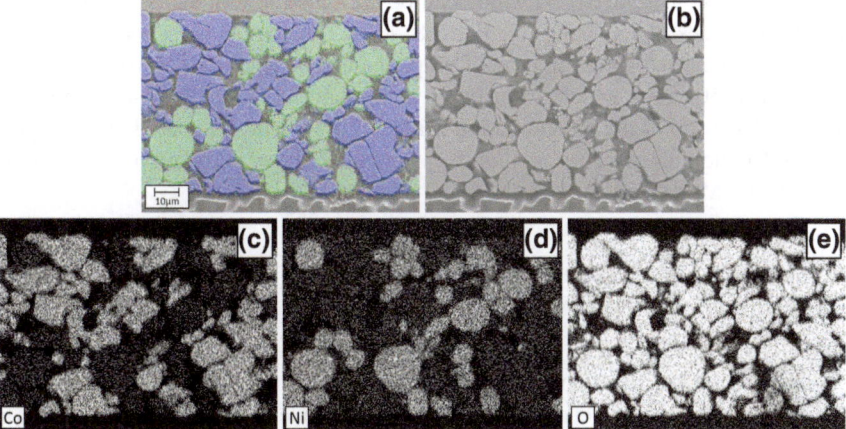

Fig. 5.7 EDX mapping of a cross section of a cathode. **a** Mix of the EDX signals. *Green* indicates $LiNi_{0.5}Mn_{0.3}Co_{0.2}O_2$, and *blue* $LiCoO_2$. **b** Electron image of the cross section. **c–e** EDX signals of the $K\alpha_1$ line of cobalt, nickel, and oxygen

5.3.4 Raman Spectroscopy

Raman spectroscopy can be used to investigate the exfoliation of graphene layers that can occur on graphite anodes. The Raman spectrum of highly ordered graphite

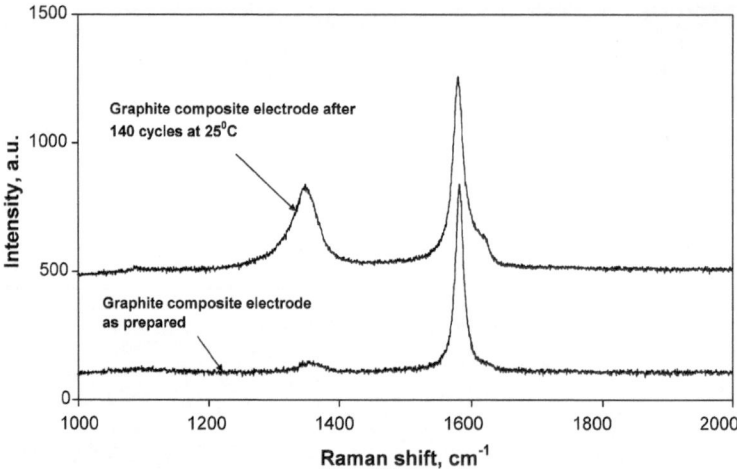

Fig. 5.8 Raman spectra of a pristine composite graphite electrode and the same electrode after 140 potentiodynamic intercalation–deintercalation cycles ($10\,\mu V\,s^{-1}$) in a 1 M LiPF$_6$ in EC:EMC 1:2 electrolyte solution at 25 °C. (Reprinted from [12] (Fig. 3), Copyright 2005, with permission from Elsevier)

shows one band at $1580\,\mathrm{cm}^{-1}$, which can be attributed to in-plane symmetric C–C stretching modes, the G band (cf. Fig. 5.8) [12]. Disorder in the crystal structure, which can be caused by exfoliation of graphene layers, leads to the evolution of the D band at $1350\,\mathrm{cm}^{-1}$, which is attributed to a breathing mode of sp^2-carbon atoms in rings. The ratio of the integrated intensities of both bands is related to the degree of disorder and thus serves as an indicator for the degree of exfoliation of the anode, as shown by Markevich et al. [12].

5.4 Analysis of Separator

Degradation of the separator has a significant influence on the performance of a lithium-ion battery. Norin et al. [14] showed the influence of the separator degradation on the power loss in LiBs that had been exposed to elevated temperatures. The ionic conductivity of all cells decreased, and atomic force microscopy (AFM) revealed a significant loss in porosity. They concluded that the loss in porosity resulted in the rise of the separator impedance and hence a loss of ion conductivity. In addition, pore clogging due to electrolyte decomposition leads to an increase of the separator and thus cell impedance [8]. Peabody et al. [15] investigated the influence of external stress and mechanical stress due to electrode expansion and contraction during charge and discharge. Mechanical stress such as pressure can cause viscoelastic creep in the separator material. This viscoelastic creep leads to closing of the pores and hence to an increase of the cell impedance, resulting in a loss of capacity and power.

Hence, analysis of the porosity, pore size and resistance of separators is necessary for the investigation of ageing effects.

SEM and AFM are image-based tools for analyzing separators. With the help of these methods, the surface and the general structure of the separators can be analyzed, as well as the distribution, structure and diameter of the pores.

For more precise investigations on the influence of these properties on the ageing of lithium-ion batteries, quantitative methods have to be applied.

One quantitative method for measuring the air permeability is the determination of the Gurley number t_G, which is dependent on the porosity (ε), pore size (d), thickness (l) and tortuosity (τ) of the membrane investigated [4, 21]:

$$t_G = 5.18 \times 10^{-3} \times \frac{\tau^2 \times l}{\varepsilon \times d} . \tag{5.1}$$

Pinholes in the separator lead to low Gurley numbers. Thus, several different samples from the same cell have to be measured to obtain reliable results.

The MacMullin number (N_m) is calculated by the ratio of the resistivity of the electrolyte-filled separator (ρ_S) and the resistivity of the electrolyte alone (ρ_E) [1, 11]:

$$N_m = \frac{\rho_S}{\rho_E} = \frac{\tau^2}{\varepsilon} \tag{5.2}$$

with

$$\rho_x = \frac{R_x \times A}{l} \tag{5.3}$$

where τ is the tortuosity, ε is the porosity, R_x is the measured resistance, A is the area of the sample, and l is the thickness of the membrane. Determination of the electrical resistance of a membrane is considered to be a more accurate way to measure the permeability of a membrane than air permeability [21].

Mercury intrusion porosimetry is used to determine the porosity of pristine and aged separators. Together with the MacMullin number the tortuosity of the membranes can be calculated.

5.5 Ageing of Electrolytes

To obtain a full image on the ageing behavior of an organic electrolyte used in an LiB, it is necessary to combine several analytical methods. There is no method available that can simultaneously detect ionic compounds and organic compounds, as well as elemental impurities. Combining all of the methods presented in this section leads to the decomposition scheme of electrolyte and the containing conducting salt (Sect. 5.6).

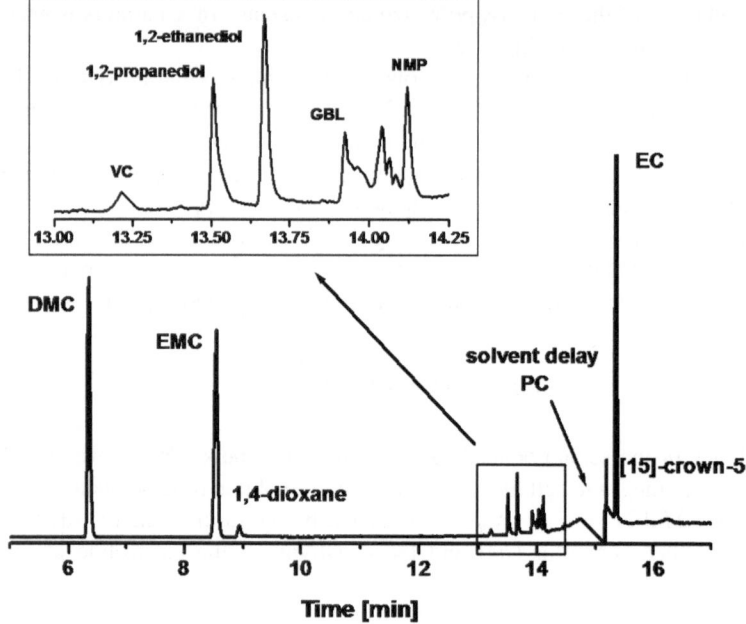

Fig. 5.9 GC-MS Chromatogram of an anode washing solution of a commercially available electrode. (Reprinted from [25] (Fig. 4), Copyright 2014, with permission from Elsevier)

5.5.1 Gas Chromatography (GC)

The volatile species in the electrolyte are usually detected and identified with the help of gas chromatography. The detector is normally a mass spectrometer (MS) or a flame ionization detector (FID), and both are sometimes used simultaneously. Depending on the composition and the target compounds, detectors such as the nitrogen-phosphorous detector (NPD), the thermal conductivity detector (TCD) or the electron capture detector (ECD) can be used in addition. Furthermore, using the headspace sampling [25] technique, electrolyte residues which stick to the surfaces or are incorporated into the electrodes can be identified and quantified using standards. Compared to the other detectors, the mass spectrometer is normally used to identify compounds, but it can also be used for quantification. A GC-MS chromatogram of an anode washing solution is shown in Fig. 5.9. A large variety of different compounds were detected.

5.5.2 Ion Chromatography (IC)

Ion chromatography can be used in different ways in electrolyte investigations. First, this method can be used to determine the elemental concentrations of halides

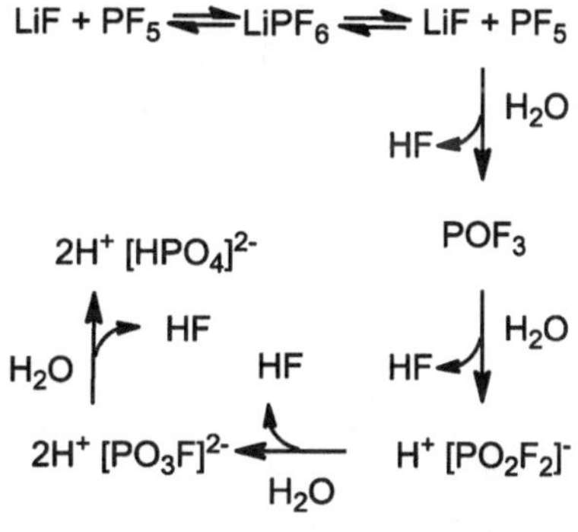

Fig. 5.10 Proposed decomposition pathway of lithium hexafluorophosphate (Reprinted from [23] (Fig. 6), Copyright 2012, with permission from Elsevier)

$$LiF + PF_5 \rightleftharpoons LiPF_6 \rightleftharpoons LiF + PF_5$$

$$HF \longleftarrow \Big\downarrow^{H_2O}$$

$$2H^+ [HPO_4]^{2-} \qquad\qquad POF_3$$

$$H_2O \Big\uparrow \hspace{-0.5em} \searrow^{HF} \qquad HF \qquad HF \longleftarrow \Big\downarrow^{H_2O}$$

$$2H^+ [PO_3F]^{2-} \longleftarrow \hspace{-1em}\underset{H_2O}{} H^+ [PO_2F_2]^-$$

(especially fluorine), alkali and alkaline earth metals. Second, in combination with electrospray ionization mass spectrometry (ESI-MS), IC can identify unknown decomposition products via their mass-to-charge (m/z) ratio. Therefore, it is a very important tool for the characterization of ageing effects in all kinds of electrolytes [10, 24].

5.5.3 ICP-OES / TXRF

These techniques are used to detect elemental impurities in the electrolyte. In comparison to the IC, these techniques are particularly well-suited for quantifying metal contents. Depending on the exact composition of the samples, radial ICP-OES systems are preferred over axial systems, since the special setup allows for the analysis of a higher organic content without additional dilution steps. With matrix matching standards, even quantitative information is accessible.

Furthermore, the coupling of IC and ICP-OES or ICP-MS can deliver valuable quantitative information when there are no standards available, especially in the case of the decomposition products. TXRF requires only small amounts of electrolyte (5–10 μL), but it cannot measure lithium. Therefore, it is used for the determination of transition metal impurities.

5.6 Decomposition Scheme of Commercial Electrolyte

Combining all of the methods described above makes it possible to develop a decomposition scheme of the electrolyte under investigation. Figure 5.10 shows the decomposition scheme of the conducting salt $LiPF_6$.

The decomposition scheme of a standard electrolyte based on a mixture of ethylene carbonate and diethyl carbonate is shown in Fig. 5.11.

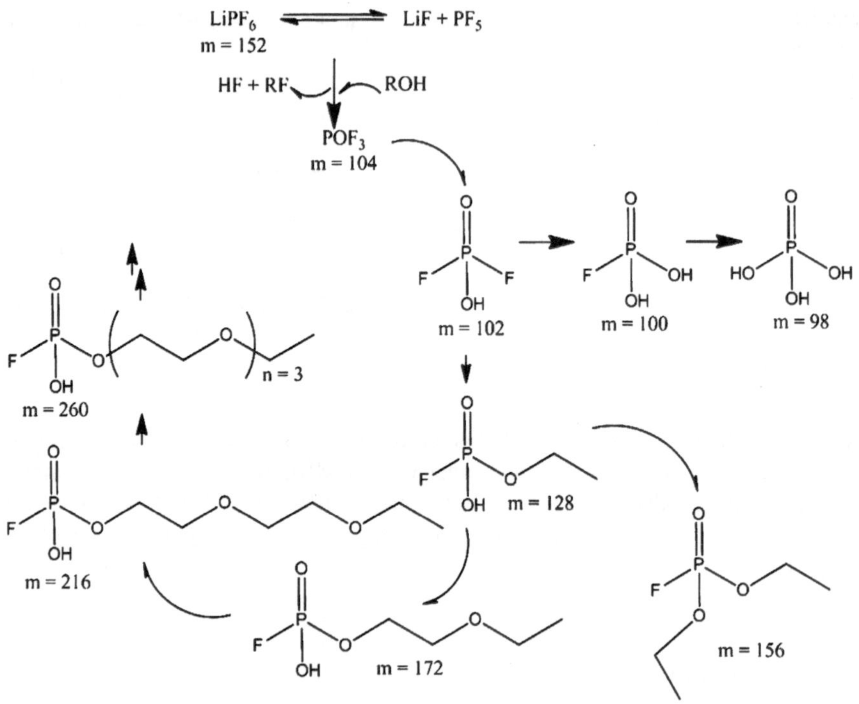

Fig. 5.11 Proposed decomposition pathway of EC:DEC 3/7 (Reprinted from [24] (Fig. 7), Copyright 2013, with permission from Elsevier)

5.7 Quantitative Measurements

The need for quantitative measurements and results regarding the ageing effects in lithium-ion batteries is quite high. Since quantitative results always depend on the calibration and therefore on the availability and quality of certified standards or reference materials, it is difficult to quantify new/unknown decomposition products or compounds which are not stable as a solution.

For elemental analyses such as ICP-OES, TXRF or, in some cases, IC (e.g. the halides), the situation is quite easy. For every element, there is a single or multielement standard available. But since the standards are normally aqueous, the matrix of the electrolyte or the dissolved electrode (e.g. acidic content) has to be matched. Otherwise, incorrect results will be obtained, which leads to the drawing of wrong conclusions regarding the ageing effects. One exception to the rule is the LA-ICP-MS, since the available standards normally consist of a glass matrix, which does not match the electrodes. Therefore, new standards have to be developed to gain quantitative information with this method.

For most of the organic carbonates and additives, the situation is the same. With the matching standards and matrix, the compounds can be quantified via GC or

HPLC measurements. When there is no standard available, especially for coupling techniques such as IC-ESI/MS or LC-ESI/MS, the concentration cannot be quantified or even estimated. One approach is the coupling of the chromatographic system to an ICP-OES or ICP-MS system. With this combination and a compound which is similar to the target compounds it is possible to obtain quantitative information.

In general, it can be concluded that, besides the development of new techniques and methods for the investigation of ageing effects in lithium-ion batteries, new standard materials and certified reference materials have to be developed as well. Otherwise, the exact influence of ageing effects cannot be quantitatively derived from measurements.

Acknowledgments The authors acknowledge the financial support of Westfälische Wilhelms-Universität Münster, the Ministry for Economic Affairs, Energy and Industry (MWEIHM) of the State of North Rhine-Westphalia, the Ministry of Innovation, Science and Research (MIWF) of the State of North Rhine-Westphalia, the German Federal Ministry of Economics and Technology (BMWi), and the German Federal Ministry of Education and Research (BMBF). Sascha Weber furthermore acknowledges the financial support of the "COMET K2 - Competence Centers for Excellent Technologies Program" of the Austrian Federal Ministry for Transport, Innovation and Technology (BMVIT), the Austrian Federal Ministry of Economy, Family and Youth (BMWFJ), the Austrian Research Promotion Agency (FFG), the Province of Styria, and the Styrian Business Promotion Agency (SFG).

References

1. Arora P, Zhang ZJ (2004) Battery separators. Chem Rev 104(10):4419–4462. doi:10.1021/cr020738u
2. Barré A, Deguilhem B, Grolleau S, Gérard M, Suard F, Riu D (2013) A review on lithium-ion battery ageing mechanisms and estimations for automotive applications. J Power Sources 241:680–689. doi:10.1016/j.jpowsour.2013.05.040
3. Besenhard JO, Winter M, Yang J, Biberacher W (1995) Filming mechanism of lithium-carbon anodes in organic and inorganic electrolytes. J Power Sources 54(2):228–231. doi:10.1016/0378-7753(94)02073-C
4. Callahan RW, Nguyen KV, McLean JG, Propost J, Hoffman DK (1993) In: Proceedings of the 10th international seminar on primary and secondary battery technology and application, Fort Lauderdale, Florida
5. Goldstein J, Newbury DE, Joy DC, Lyman CE, Echlin P, Lifshin E, Sawyer L, Michael JR (2003) Scanning electron microscopy and X-ray microanalysis, 3rd edn. Springer, New York
6. Grützke M, Mönnighoff X, Winter M, Nowak S (2013) Extraction of organic carbonate based electrolytes with supercritical carbon dioxide for a high efficient recycling of lithium-ion batteries. In: 224th ECS meeting, https://ecs.confex.com/ecs/224/webprogram/Paper22925.html. Accessed 10 Sept 2013
7. Kohs W, Santner HJ, Hofer F, Schröttner H, Doninger J, Barsukov I, Buqa H, Albering JH, Möller KC, Besenhard JO, Winter M (2003) A study on electrolyte interactions with graphite anodes exhibiting structures with various amounts of rhombohedral phase. J Power Sources 119–121:528–537. doi:10.1016/S0378-7753(03)00278-7
8. Kostecki R, Norin L, Song X, McLarnon F (2004) Diagnostic studies of polyolefin separators in high-power Li-ion cells. J Electrochem Soc 151(4):A522–A526. doi:10.1149/1.1649233

9. Krämer E, Schmitz R, Niehoff P, Passerini S, Winter M (2012) SEI-forming mechanism of 1-Fluoropropane-2-one in lithium-ion batteries. Electrochim Acta 81:161–165. doi:10.1016/j. electacta.2012.07.091

10. Lux SF, Terborg L, Hachmöller O, Placke T, Meyer HW, Passerini S, Winter M, Nowak S (2013) LiTFSI stability in water and its possible use in aqueous lithium-ion batteries: pH dependency, electrochemical window and temperature stability. J Electrochem Soc 160(10):A1694–A1700. doi:10.1149/2.039310jes

11. MacMullin RB, Muccini GA (1956) Characteristics of porous beds and structures. AIChE J 2(3):393–403. doi:10.1002/aic.690020320

12. Markevich E, Salitra G, Levi MD, Aurbach D (2005) Capacity fading of lithiated graphite electrodes studied by a combination of electroanalytical methods, Raman spectroscopy and SEM. J Power Sources 146(1–2):146–150. doi:10.1016/j.jpowsour.2005.03.107

13. Niehoff P, Passerini S, Winter M (2013) Interface investigations of a commercial lithium ion battery graphite anode material by sputter depth profile X-ray photoelectron spectroscopy. Langmuir 29(19):5806–5816. doi:10.1021/la400764r

14. Norin L, Kostecki R, McLarnon F (2002) Study of membrane degradation in high-power lithium-ion cells. Electrochem Solid-State Lett 5(4):A67–A69. doi:10.1149/1.1457206

15. Peabody C, Arnold CB (2011) The role of mechanically induced separator creep in lithium-ion battery capacity fade. J Power Sources 196(19):8147–8153. doi:10.1016/j.jpowsour.2011.05.023

16. Peled E (1979) The electrochemical behavior of alkali and alkaline earth metals in nonaqueous battery systems-the solid electrolyte interphase model. J Electrochem Soc 126(12):2047–2051. doi:10.1149/1.2128859

17. Prochazka W, Pregartner G, Cifrain M (2013) Design-of-experiment and statistical modeling of a large scale aging experiment for two popular lithium ion cell chemistries. J Electrochem Soc 160(8):A1039–A1051. doi:10.1149/2.003308jes

18. Santner HJ, Möller KC, Ivančo J, Ramsey MG, Netzer FP, Yamaguchi S, Besenhard JO, Winter M (2003) Acrylic acid nitrile, a film-forming electrolyte component for lithium-ion batteries, which belongs to the family of additives containing vinyl groups. J Power Sources 119–121:368–372. doi:10.1016/S0378-7753(03)00268-4

19. Santner HJ, Korepp C, Winter M, Besenhard JO, Möller KC (2004) In-situ FTIR investigations on the reduction of vinylene electrolyte additives suitable for use in lithium-ion batteries. Anal Bioanal Chem 379(2):266–271. doi:10.1007/s00216-004-2522-4

20. Song KM, Park SW, Hong WH, Lee H, Kwak SS, Liu JR (1992) Isolation of Vindoline from Catharanthus roseus by supercritical fluid extraction. Biotechnol Prog 8(6):583–586. doi:10.1021/bp00018a018

21. Spotnitz R (2011) Separators for lithium-ion batteries. In: Daniel C, Besenhard J (eds) Handbook of battery materials, 2nd edn. Wiley-VCH, Weinheim, chap 20, doi:10.1002/9783527637188.ch20

22. Stashenko EE, Puertas MA, Combariza MY (1996) Volatile secondary metabolites from Spilanthes americana obtained by simultaneous steam distillation-solvent extraction and supercritical fluid extraction. J Chromatogr A 752(1–2):223–232. doi:10.1016/S0021-9673(96)00480-3

23. Terborg L, Nowak S, Passerini S, Winter M, Karst U, Haddad PR, Nesterenko PN (2012) Ion chromatographic determination of hydrolysis products of hexafluorophosphate salts in aqueous solution. Anal Chim Acta 714:121–126. doi:10.1016/j.aca.2011.11.056

24. Terborg L, Weber S, Blaske F, Passerini S, Winter M, Karst U, Nowak S (2013) Investigation of thermal aging and hydrolysis mechanisms in commercial lithium ion battery electrolyte. J Power Sources 242:832–837. doi:10.1016/j.jpowsour.2013.05.125

25. Terborg L, Weber S, Passerini S, Winter M, Karst U, Nowak S (2014) Development of gas chromatographic methods for the analyses of organic carbonate-based electrolytes. J Power Sources 245:836–840. doi:10.1016/j.jpowsour.2013.07.030

26. Vetter J, Novák P, Wagner MR, Veit C, Möller KC, Besenhard JO, Winter M, Wohlfahrt-Mehrens M, Vogler C, Hammouche A (2005) Ageing mechanisms in lithium-ion batteries. J Power Sources 147(1–2):269–281. doi:10.1016/j.jpowsour.2005.01.006

27. Wagner MR, Raimann PR, Trifonova A, Moeller KC, Besenhard JO, Winter M (2004) Electrolyte decomposition reactions on tin- and graphite-based anodes are different. Electrochem Solid-State Lett 7(7):A201–A206. doi:10.1149/1.1739312

28. Winter M (2009) The solid electrolyte interphase - the most important and the least understood solid electrolyte in rechargeable Li batteries. Z Phys Chem 223(10–11):1395–1406. doi:10.1524/zpch.2009.6086

29. Winter M, Besenhard JO (1999) Wiederaufladbare Batterien. Chem unserer Zeit 33(6):320–332. doi:10.1002/ciuz.19990330603

30. Winter M, Brodd RJ (2004) What are batteries, fuel cells, and supercapacitors? Chem Rev 104(10):4245–4269. doi:10.1021/cr020730k

31. Winter M, Besenhard JO, Spahr ME, Novák P (1998) Insertion electrode materials for rechargeable lithium batteries. Adv Mater 10(10):725–763. doi:10.1002/(SICI)1521-4095(199807)10:10<725::AID-ADMA725>3.0.CO;2-Z

32. Winter M, Imhof R, Joho F, Novák P (1999) FTIR and DEMS investigations on the electroreduction of chloroethylene carbonate-based electrolyte solutions for lithium-ion cells. J Power Sources 81–82:818–823. doi:10.1016/S0378-7753(99)00116-0

33. Winter M, Appel WK, Evers B, Hodal T, Möller KC, Schneider I, Wachtler M, Wagner MR, Wrodnigg GH, Besenhard JO (2001) Studies on the anode/electrolyte interface in lithium ion batteries. Monatsh Chem 132(4):473–486. doi:10.1007/s007060170110

34. Yu DY, Donoue K, Kadohata T, Murata T, Matsuta S, Fujitani S (2008) Impurities in liFePO$_4$ and their influence on material characteristics. J Electrochem Soc 155(7):A526–A530. doi:10.1149/1.2919105

35. Zhang SS (2006) A review on electrolyte additives for lithium-ion batteries. J Power Sources 162(2):1379–1394. doi:10.1016/j.jpowsour.2006.07.074

36. Zhang SS, Jow TR (2002) Aluminum corrosion in electrolyte of Li-ion battery. J Power Sources 109(2):458–464. doi:10.1016/S0378-7753(02)00110-6

Chapter 6
Bayesian Inference for Lithium-Ion Cell Parameter Estimation

Matthias K. Scharrer, Heikki Haario and Daniel Watzenig

Abstract The optimization of lithium-ion cells is becoming increasingly important. Using models that reflect the fundamental electrochemical processes is advantageous for this purpose. These models are typically computationally expensive and hard to invert using optimization methods. Additionally, deterministic optimization methods do not yield information regarding parameter uncertainties in the presence of noise. To overcome this problem, it is possible to apply Bayesian methods. This chapter provides an overview of parameter estimation. After a brief introduction to the model, parameter selection and modelling of the prior is presented. Finally, we present the results of a synthetic fitting problem solved by a parallel adaptive Markov chain Monte Carlo method. We validate the approach and compare it to realistic noisy data and a separated method.

Keywords Parameter estimation · Parallel MCMC · Lithium-ion cell model

Nomenclature

A_i	Inner surface
$brugg_x$	Bruggemann coefficient
C_{dl}	Double-layer capacity
c_x	Li^+-concentration in phase x

M. K. Scharrer (✉) · D. Watzenig
Virtual Vehicle Research Center, Graz, Austria
e-mail: matthias.scharrer@v2c2.at

D. Watzenig
e-mail: daniel.watzenig@v2c2.at

H. Haario
Lappeenranta University of Technology, Lappeenranta, Finland
e-mail: heikki.haario@lut.fi

A. Thaler and D. Watzenig (eds.), *Automotive Battery Technology*,
Automotive Engineering: Simulation and Validation Methods,
DOI: 10.1007/978-3-319-02523-0_6, © The Author(s) 2014

$c_{x,0}$	Initial Li$^+$-concentration
$\hat{D}_x$	Diffusivity
D_x	Effective Diffusivity
$d_i(\boldsymbol{\theta})$	Elementary effect of variable θ_i
e	Random number in $(0,1)$
F	Faraday's constant
$f(\cdot, \boldsymbol{\theta})$	Model output
F_i	Distribution of effects of θ_i
$\bar{f}(\bar{\boldsymbol{\theta}})$	Model output for target parameters
$i(t)$	Cell current density
j_{BV}^*	Butler-Volmer current density
k	Exchange current density and reaction rate
$k_{a:c}$	Combined factor k_a/k_c
n	Number of sample
$p(y\|\boldsymbol{\theta})$	Likelihood function
$q(\cdot\|\boldsymbol{\theta}_n)$	Parameter proposal distribution
Q_1	Space-time cylinder of Ω
Q_1'	Space-time cylinder of Ω'
Q_2	Space-time cylinder of Λ
r	Radial coordinate in particles
R_0	Proposal deviation of $q(\cdot\|\theta_0)$
R_a	Particle radius in anode
R_c	Particle radius in cathode
R_g	Universal gas constant
R_n	Proposal deviation of $q(\cdot\|\theta_n)$ at sample n
T	Final simulation time
t_i	Time at step i of reference model
t_j	Time at step j of model under test
t_ℓ^+	Transference number of cations
u	Unknowns of the model
U_{OCV}	Equilibrium potential function
y, y_i	Points of measurements
z	Number of transferred electrons
α	Charge transfer coefficient
$\alpha(\cdot, \hat{\boldsymbol{\theta}})$	Acceptance probability
ε	Measurement error
ε_x	Phase x volume fraction
$\varepsilon_{\ell:s}$	Combined factor $\varepsilon_\ell/\varepsilon_s$
$\varepsilon_{\ell+s}$	Combined factor $\varepsilon_\ell + \varepsilon_s$
ε_Σ	Small value to add to R_n
φ_x	Electric potential in phase x

η	Overpotential in j_{BV}^*
$\hat{\kappa}(c\ell)$	Ionic conductivity function
$\kappa\,(c_\ell)$	Effective ionic conductivity
Λ_a	Anode domain in particles
Λ_c	Cathode domain in particles
Λ	Combined anode and cathode model domain
μ	Sample mean
μ_ℓ	Migration coefficient
Ω_a	Anode model domain
Ω_c	Cathode model domain
Ω_s	Separator model domain
Ω	Entire model domain
Ω'	Combined anode and cathode model domain
$\pi(\theta)$	Prior probability
$\pi(\theta|y)$	Posterior probability
σ	Sample or measurement noise standard deviation
$\hat{\sigma}_s$	Electronic conductivity
σ_s	Effective electronic conductivity
θ	Parameter set of interest
θ_0	Initial parameter values
θ_a	Lower parameter bounds
τ	Integrated autocorrelation time
θ_b	Upper parameter bounds
$\hat{\theta}$	Proposed parameter
$\tilde{\theta}(\theta)$	Transformed parameters
$\bar{\theta}$	Target parameters
Θ	Parameter space
ϑ	Temperature

6.1 Introduction

Compared to combustion-engine-based vehicles, battery-powered and hybrid vehicles are clearly more environmentally friendly during operation. To increase the efficiency of batteries and control strategies, having knowledge about the internal state and material parameters is becoming increasingly important. The internal state is a comprehensive term that includes abstract quantities (e.g. state of charge (SoC), state of function and state of health) and physical quantities (e.g. potentials and concentrations). The latter are reconstructed as closely as possible from measurement data in order to deduce material parameters which are not accessible or measurable.

In order to describe internal states and processes within lithium-ion cells, Fuller et al. devised a system of coupled non-linear partial differential equations (PDEs) in one dimension [2], which was subsequently improved and extended by Newman and Thomas-Alyea [5] and many others. This system models the cell in terms of transport equations for lithium ions, chemical interaction and electronic field computations in active particles of anode and cathode. These equations are coupled by modelling electrode kinetics that occur on the particle surfaces of electrodes.

Parameter estimation techniques focus on non-invasive methods only (i.e. without the need to destructively open the cell). These methods estimate parameters by matching predicted cell model output voltages for a given current profile to experimental measurements. Regarding the cell model, these studies require up to 50 parameters to be tested. To get a better insight into the model behaviour and to reduce the number of parameters being tested, a modified "Morris-one-at-a-time" (MOAT) parameter screening was conducted [7].

Using the information obtained from parameter screening, a subset of relevant parameters was selected. Taking prior knowledge about this subset into account, we fed this information into a Markov chain Monte Carlo (MCMC) method.

Here, we demonstrate the application of MCMC for three cases using a recent parallel adaptive sampling approach: estimate parameters and their uncertainty using a synthetic profile without noise, estimate uncertainty in the presence of noise and estimate uncertainty in the presence of noise but without exploiting the power of parallelism.

The remainder of this chapter is structured as follows: Sect. 6.2 provides an introduction to parameter estimation in general. Section 6.3 defines the simulation framework of the cell model and briefly summarizes the solution procedure, Sect. 6.4 outlines the parameter screening method using MOAT as a way to reduce the complexity of the parameter estimation problem. In Sect. 6.5 the inversion problem and influences regarding measurement error are discussed. Finally, the solution algorithm for estimating the parameters is defined using statistical methods.

6.2 Inverse Problems: Making the Invisible Visible

6.2.1 Introduction

The aim of measurements is to get information about the system or phenomenon under study. However, it is quite typical that we cannot directly measure quantities of interest. Rather, the measured data depends, in some way, on the quantities desired, and so at least contains some information about them. The relationship between the quantities and data is described by a mathematical model. Solving the mathematical model—calculating the output of it with certain inputs—is called a *forward* problem. However, in order to obtain correct model predictions, the model must be correctly formulated and calibrated. Starting with the data that we have measured, the problem of trying to reconstruct the quantities that we actually want is called an *inverse problem*. Loosely speaking, we say that an inverse problem is where we measure an *effect* and want to determine the *cause*.

Most science is data-driven in this way. Indeed, inverse problems are ubiquitous in nature. All our senses, and the senses of any animals, are perfect examples of often amazing solutions to inverse problems. Bats find their way in complete darkness by emitting sonar waves and constructing their environment by received echoes. Whales and dolphins do the same in the ocean over long distances, and they can also locate prey and predators. Our brain is able to produce an on-line image of our environment by the complex scattered light waves received by our eyes. Inverse problem research is progressing in various fields of science by making the 'invisible to visible', although we are still far behind nature in many respects. Most typical operational examples are various non-invasive imaging techniques in medicine (e.g. ultra sound, X-ray, dental tomography) and similar non-destructive methods in engineering, such as electromagnetic sounding in ore prospecting, imaging techniques in security checking, or timber tomography.

A somewhat more restricted type of data-driven problem is parameter estimation. Here, we are not constructing continuous images of any media, but rather want to calibrate a few given parameters of a mathematical model using measured data. To recall the standard setting, consider a non-linear model

$$\mathbf{y} = f(\mathbf{x}, \theta) + \varepsilon, \tag{6.1}$$

where $\mathbf{y}$ are the obtained measurements, $f(\mathbf{x}, \theta)$ is the model with design variables $\mathbf{x}$ and unknown parameters θ, and the measurement error is denoted by ε. The formulation of the model f is typically based on the first principles of physics and chemistry, while solving the forward problem may require advanced numerical methods, as well as effective computers. The most common approach for estimating the values of the parameters is to match them against measurements using the Least Squares, LSQ, criterion: minimize the sum of squares

$$l(\theta) = \sum_{i=1}^{n} [y_i - f(x_i, \theta)]^2. \tag{6.2}$$

In practice, for most models one can use standard optimization routines implemented in computational software packages. However, since all available data contains measurement errors, the estimated unknowns are to some degree uncertain. A natural question then arises: if measurement noise corrupting the data follows some statistical distribution, what is the *distribution* of the possible solutions after the estimation procedure? This question is the core of the study of *statistical inverse problems*. We will discuss the Bayesian (probabilistic) framework below to give practical answers to this question.

Noisy data is, however, not the only source of uncertainty in modelling. It may be more challenging to estimate the impact of model bias due to insufficient understanding of the phenomena under study, or just the numerical approximation errors we may have to make to minimize the CPU requirements. A recent development is the focus on *uncertainty quantification* (UQ) within computational models, particularly in the computational science and engineering community.

6.2.2 Deterministic Approaches: Linear and Linearized Models

For linear models, the statistics of parameter estimation can be readily determined. Consider a linear model with p variables, $f(\mathbf{x}, \theta) = \theta_0 + \theta_1 x_1 + \theta_2 x_2 + \ldots + \theta_p \mathbf{x}_p$, with noisy measurements $\mathbf{y} = (y_1, y_2, \ldots, y_n)$ obtained at points $\mathbf{x}_i = (x_{1i}, x_{2i}, \ldots, x_{ni})$ where $i = 1, \ldots, p$. We can write the model in matrix notation,

$$\mathbf{y} = \mathbf{X}\theta + \varepsilon, \tag{6.3}$$

where $\mathbf{X}$ is the design matrix that contains the measured values for the control variables, augmented with a column of ones to account for the intercept term θ_0. It is not difficult to derive a direct formula for the LSQ estimator that minimizes $SS(\theta) = \|\mathbf{y} - \mathbf{X}\theta\|_2^2$. The solution to the *normal equations* $\mathbf{X}^T\mathbf{X}\theta = \mathbf{X}^T\mathbf{y}$ is written as

$$\hat{\theta} = (\mathbf{X}^T\mathbf{X})^{-1}\mathbf{X}^T\mathbf{y}. \tag{6.4}$$

To obtain the statistics for the estimate, we can compute the covariance matrix $\text{Cov}(\hat{\theta})$. With the common assumption that there is independent and identically distributed measurement noise with measurement error variance σ^2, the covariance matrix for the measurement is given as $\text{Cov}(\mathbf{y}) = \sigma^2\mathbf{I}$, where $\mathbf{I}$ is the identity matrix. The parameter uncertainty is then characterized by the covariance

$$\text{Cov}(\hat{\theta}) = \sigma^2(\mathbf{X}^T\mathbf{X})^{-1}. \tag{6.5}$$

The diagonal elements of the covariance matrix give the variances of the estimated parameters, which are often reported by different statistical software packages. If we further assume that the measurement errors are Gaussian, we can also conclude that the distribution of $\hat{\theta}$ is Gaussian, with the mean and covariance matrix given by the formulae above.

For non-linear models, no such direct formulae are available, and one must resort to numerical methods and different approximations. The standard strategy in such cases is to linearize the non-linear model and simply employ the linear theory. This leads to the computation of the derivatives of the model with respect to the parameters. The first derivatives can be collected into a matrix that is called the Jacobian matrix $\mathbf{J}$, which has elements

$$[\mathbf{J}]_{ip} = \frac{\partial f(x_i; \theta)}{\partial \theta_p}\Big|_{\theta=\hat{\theta}}, \tag{6.6}$$

where the notation indicates that the derivatives are evaluated at the estimate $\hat{\theta}$, at every measurement point x_i. The Jacobian matrix $\mathbf{J}$ assumes the role of the design matrix $\mathbf{X}$ in the linear case. That is, the approximative error analysis for non-linear models, assuming independent and identically distributed Gaussian errors with measurement error variance σ^2, is given by the covariance matrix

$$\text{Cov}(\hat{\theta}) = \sigma^2(\mathbf{J}^T\mathbf{J})^{-1}. \tag{6.7}$$

The measurement error σ^2 can be estimated using repeated measurements. Often, however, replicated measurements are not available. In this case, the measurement noise can be estimated using the residuals of the fit, using the 'perfect model' assumption that *residuals* $\approx$ *measurement error*. An estimate for the measurement error can be obtained using the mean square error (MSE):

$$\sigma^2 \approx MSE = RSS/(n - p), \tag{6.8}$$

where RSS (residual sum of squares) is the fitted value of the least squares function, n is the number of measurements, and p is the number of parameters. That is, the measurement error is computed as the average of the squared residuals, corrected by the number of estimated parameters.

6.2.3 Bayesian Methodology

The approximative error estimates available from the linearisations are satisfactory only if the true parameter distributions are close to the Gaussian distributions that the linerarisation produces. However, there is no guarantee on this; often the non-linear dependencies yield distributions that are far from the linearized ones. Here, we discuss methods that reveal the true distributions, even with strong non-linearities. We will show how these approaches, based on random or "Monte Carlo" calculations, are able to produce samples from underlying distributions that converge to the true distributions, if the sample size is large enough. Indeed, the sampling algorithms are able to solve a "mission impossible": we create an approximation of an unknown distribution by sampling from it—even if we do not know from where to sample!

Samples may be produced by iterating the parameter estimation process several times. To get different results, we must randomize some part of the process. There are basically to options:

- Perturb the measured data and refit the parameters.
- Perturb parameters, accept parameter values that give good enough fits to data, reject others.

Both options are based on the fact that data contains randomness or noise: in the experiment, we might equally well have obtained somewhat different data points, and thus different parameter estimates. In the first option, we directly produce different data values. This leads to various forms of *Bootstrap* methods. In the second approach, data is not changed, but the uncertainty of data is taken into account by accepting, roughly speaking, parameters that produce model predictions that fit the data within the noise level of measurements. This approach is the background idea of several *Markov chain Monte Carlo*, or MCMC methods.

Both approaches might seem intuitively appealing. However, different interpretations arise, which have been the focus of a long-standing dispute between two opposing views on statistical methods, the "Frequentists against the Bayesians". A key question here is how we interpret the nature of a parameter to be estimated: is it a constant or variable? Treating parameters as constants is known as the *classical* or *frequentist* perspective in statistics. Bayesian thinking explicitly allows for the unknown vector of parameters θ to be interpreted as a *random variable* with a distribution of its own. This reflects the belief that θ is not a fixed vector, but rather can vary. In addition, this approach typically emphasizes the use of *prior knowledge*, even of a subjective nature, in the estimation process.

A practically oriented researcher might find the dispute somewhat academic. In a real modelling project, are we so concerned about the "true" interpretation of parameters? However, we certainly should be interested in the *reliability* of the model predictions. It is essential to realize that parameter estimation problems do not have a unique solution. A multitude of different parameter combinations may fit the data "equally well", from the statistical point of view, as we take into account the noise in the measurements.

6.2.4 Markov Chain Monte Carlo: MCMC

In general, MCMC methods keep the measured data intact, but vary the parameters to find "all" solutions close to measurements. Therefore, they belong to the Bayesian school of methods. Indeed, for many scientists, the Bayesian approach is almost synonymous with the use of MCMC methods. Moreover, the Bayesian methods have gained tremendously in popularity in the last 20 years because of new MCMC algorithms: they provide solutions to problems that classical methods simply cannot handle. Another obvious reason for this trend is the continued evolution of more powerful computers that can successfully run the algorithms.

There are several advantages of using MCMC to solve parameter estimation or "inverse" problems. First, full characterization of (non-Gaussian) posterior distributions is possible. Second, we have full freedom in implementing prior information. Even modelling errors can be taken into account in a flexible way. Moreover, we are less likely to get trapped in local minimums than when employing optimization methods to get maximum *a posteriori* (MAP) estimates.

To fix ideas, consider again the generic non-linear model $y_i = f(x_i; \theta) + \varepsilon_i$, where we assume the errors are normally distributed and independent between measurements: $\varepsilon_i \overset{iid}{\sim} \mathcal{N}(0, \sigma^2)$. The vector θ is the unknown to be estimated by the measured data values y_i. However, suppose first that the true value of θ is known, and that we have a perfect, unbiased model $f(\mathbf{x}; \theta)$ to represent the true behaviour of our system. A measured value y_i will then follow the normal distribution, with $f(x_i, \theta)$ as the centre point. Since the errors ε_i were assumed to be independent, the distributions of the different measurements y_i, $i = 1, 2, \ldots n$ are independent, and the joint distribution for the measurement vector $\mathbf{y} = (y_1, \ldots y_n)$ is obtained as the product of distributions:

$$p(y|\boldsymbol{\theta}) = \prod_{i=1}^{n} \frac{1}{\sqrt{2\pi\sigma^2}} \exp\left\{-(y_i - f(x;\boldsymbol{\theta}))^2/2\sigma^2\right\} \tag{6.9}$$

$$= \frac{1}{(2\pi\sigma^2)^{n/2}} \exp\left\{-\sum_{i=1}^{n}(y_i - f(x;\boldsymbol{\theta}))^2/2\sigma^2\right\}. \tag{6.10}$$

Thus, we arrive at the familiar least squares function in the exponent, divided by the assumed measurement error variance σ^2.

Note how we write the above distribution in the form $p(y|\boldsymbol{\theta})$, which typically is used for a *conditional probability*. Indeed, as we actually do not know $\boldsymbol{\theta}$, we should interpret the above formula as the probability of y, on the condition that the true value of the unknown parameter is $\boldsymbol{\theta}$. This expression is called the *likelihood function*; it gives the likelihood of observing the measurement, with a given value for $\boldsymbol{\theta}$.

We return now to our original goal: to estimate the parameters $\boldsymbol{\theta}$ for given values of y. For this purpose, we might just change the roles of y and $\boldsymbol{\theta}$ when considering the expression (6.10) in the equation above. We could interpret it as representing the distribution of $\boldsymbol{\theta}$ with a given y. Note that maximizing the likelihood function with respect to $\boldsymbol{\theta}$ is equivalent to minimizing the sum of the squared residuals, so this *maximum likelihood estimator* is just the least squares solution $\widehat{\boldsymbol{\theta}}$, in this case. Moreover, values of θ that bring the values $f(x_i, \boldsymbol{\theta})$ close to data points y_i should be accepted to belong to the distribution of probable values of $\boldsymbol{\theta}$. However, we face a problem here: there is no direct way to tell exactly which values of $\boldsymbol{\theta}$ would belong to that distribution, e.g., to the region that would contain the 95 % probability mass around the maximum likelihood point. Indeed, as a function of $\boldsymbol{\theta}$, the expression (6.10) is not even a normalized probability distribution.

The Bayes formula can be written as a generalization of the conditional probability in basic probability calculus, as

$$\pi(\boldsymbol{\theta}|y) = \frac{p(y|\boldsymbol{\theta})\pi(\boldsymbol{\theta})}{\int p(y|\boldsymbol{\theta})\pi(\boldsymbol{\theta})d\boldsymbol{\theta}}. \tag{6.11}$$

We summarize the main concepts of the Bayesian analysis as follows. First, we select a *prior distribution* for the parameter vector $\boldsymbol{\theta}$, i.e. $\pi(\boldsymbol{\theta})$. We assume that the response variable y has a distribution given by the *likelihood* $p(y|\boldsymbol{\theta})$. Once we collect the data $y = (y_1, \ldots, y_n)$, we update our prior distribution for $\boldsymbol{\theta}$ using the Bayes rule to arrive at the *posterior distribution* for $\boldsymbol{\theta}$, $\pi(\boldsymbol{\theta}|y)$). To get a proper probability distribution of total mass one, we have to calculate the normalizing constant $\int p(y|\boldsymbol{\theta})\pi(\boldsymbol{\theta})d\boldsymbol{\theta}$.

In principle, the Bayes formula solves the estimation problem in a fully probabilistic sense: we find the peak, the MAP point, of the parameter distribution. Then we determine a required portion of the probability mass (typically 95 % or 99 % of the mass) around it. However, we face the problem of how to calculate the integral in the expression for the normalizing constant. The integration of the normalizing constant is often a formidable task, even for an only moderately high number of parameters in a non-linear model. Therefore, a direct application of the Bayes formula is intractable for all but trivial non-linear cases.

The MCMC methods provide a tool for handling this problem. They generate a sequence of parameter values $\theta_1, \theta_2, \ldots \theta_N$, whose empirical distribution approximates the true posterior distribution for a sufficiently large sample size N.

The trick here is that we do not know the distribution from which to sample, but we can still generate samples from it. Instead of sampling from the true distribution, we can only sample from an artificial *proposal* distribution. By combining the sampling with a simple accept/reject procedure, the posterior can be correctly approximated.

One of the most widely used MCMC algorithms is the random walk Metropolis algorithm, which appeared as early as the 1950s in statistical physics literature. The Metropolis algorithm is very simple: it works by generating candidate parameter values from a *proposal distribution* and then either accepting or rejecting the proposed value according to a simple rule. The Metropolis algorithm can be written as follows:

1. Initialize by choosing a starting point θ_1
2. Choose a new candidate $\hat{\theta}$ from a suitable **proposal distribution** $q(.|\theta_n)$, which may depend on the previous point of the chain.
3. **Accept** the candidate with probability

$$\alpha(\theta_n, \hat{\theta}) = \min\left(1, \frac{\pi(\hat{\theta})}{\pi(\theta_n)}\right). \tag{6.12}$$

If rejected, repeat the previous point in the chain. Go back to step 2.

The Metropolis algorithm assumes a symmetric proposal distribution q, i.e. the probability density of moving from the current point to the proposed point is the same as moving backwards from the proposed point to the current point: $q(\hat{\theta}|\theta_n) = q(\theta_n|\hat{\theta})$.

One can see that in the Metropolis algorithm the candidate points that give a better posterior density value than the previous point (points where $\pi(\hat{\theta}) > \pi(\theta_n)$), or move 'upward' in the posterior density function, are always accepted. However, moves 'downward' may also be accepted, with the probability given by the ratio of the posterior density values at the previous point and the proposed point. Note that only the ratios of π at consecutive points are needed, so the main difficulty is removed: the calculation of the normalizing constant is not needed, since the constant cancels out!

Most often, we may not actually want to specify a non-trivial prior distribution for the solution. We may just know that the solution components must have some bounded and positive values, leading to *uninformative* or *flat* priors. For a given parametrization then, we may just set a box of 'simple bounds', just lower and upper bounds, to constrain the solutions. The analysis is now fully driven by data, assuming that the posterior distribution of the parameters is well inside the given bounds. If, on the other hand, the posterior does not stay inside any reasonable bounds, we observe that the available data is *not sufficient to identify the parameters*. We can then consider a few options, before restricting solutions with informative priors:

- *Design of Experiments*. If non-identifiability of parameters is due to lack of data, one obvious remedy is to design new experiments to gain more informative measurements.

- *Model reductions.* Often, however, the non-identifiability is an inherent feature of the forward map, and no measurable data (or just re-parametrisation) can correct the situation. One alternative option for fixing priors for unidentified parameters is to simplify the model, and thus reduce the list of parameters to be identified.

Typically, the Bayes formula is seen as the way of putting together a *fixed* prior, data, and model. However, we can also use Bayesian sampling techniques as flexible tools in the context of design of experiments, as well as model reductions. As such, we can employ the sampling methods for *model development*, which may guide all the relevant steps of model building.

6.3 Model of a Lithium-Ion Cell

There are a wide range of different existing models for describing the performance of lithium-ion cells—see Chap. 4. Since many models are unable to reflect effects of physical processes inside the cell, we use a mechanistic model throughout the rest of the paper that defines our *forward* problem. Using the model for estimation of parameters based on observed values is then considered as our *inverse* problem

In order to describe the internal dynamic processes in a lithium-ion cell mathematically, an electrochemical model has been implemented following the widely used Doyle- Fuller-Newman (DFN) approach, [5]. Due to complex geometries at nanometre scale, together with a cell thickness of hundreds of microns, these models cannot be solved with the greatest detail. Thus, the model can be stated as a system of coupled non-linear partial differential equations in one dimension, trading off computational speed and complexity.

A lithium-ion cell with two porous intercalation electrodes (cathode in Ω_c and anode in Ω_a) and an electronically isolating separator in Ω_s in between is considered. For homogenization purposes, each electrode is assumed to consist of two phases. The solid phase is assumed to be spherical particles in both cathode (in Λ_c) and anode (in Λ_a), which line up continuously in the x direction. The liquid phase modelled in each electrode is electrolyte. In the separator Ω_s, we only consider electrolyte, as the solid phase in the separator does not participate in any reactions taken into account. Figure 6.1 shows a schematic view of the modelled domain.

The governing equations of the one-dimensional cell model considered are defined by system (6.13).

$$
\begin{aligned}
-\nabla \cdot (\sigma_s \nabla \varphi_s) &= -A_i j_{BV}^* \quad &\text{in } Q_1' \coloneqq \Omega' \times [0, T] \\
-\nabla \cdot \left(\kappa_\ell(c_\ell) \nabla \varphi_\ell + \tfrac{R_g \vartheta}{zF} \kappa_\ell(c_\ell) t_\ell^+ \tfrac{1}{c_\ell} \nabla c_\ell \right) &= A_i j_{BV}^* \quad &\text{in } Q_1 \coloneqq \Omega \times [0, T] \\
\tfrac{\partial (\varepsilon_\ell c_\ell)}{\partial t} - \nabla \cdot \left(D_\ell \left(\nabla c_\ell + \tfrac{zF}{R_g \vartheta} \mu_\ell c_\ell \nabla \varphi_\ell \right) \right) &= \tfrac{A_i}{zF} j_{BV}^* \quad &\text{in } Q_1 \\
\tfrac{\partial c_s}{\partial t} - \tfrac{1}{r^2} \tfrac{\partial}{\partial r} \left(D_s r^2 \tfrac{\partial c_s}{\partial r} \right) &= 0 \quad &\text{in } Q_2 \coloneqq \Lambda \times [0, T]
\end{aligned}
$$

$$(6.13)$$

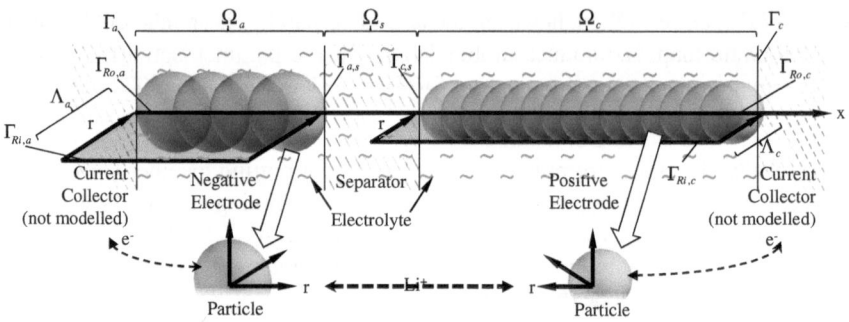

Fig. 6.1 Problem Domain: The spatial domains are defined as $\Omega = \Omega_a \cup \Omega_s \cup \Omega_c \subset \mathbb{R}$, $\Omega' = \Omega_a \cup \Omega_c$, $\Lambda_a = \Omega_a \times [0, R_a] \subset \mathbb{R}^2$, $\Lambda_c = \Omega_c \times [0, R_c] \subset \mathbb{R}^2$, $\Lambda = \Lambda_a \cup \Lambda_c$ and $R_a, R_c \in \mathbb{R}$

In the liquid phase, the variables are potentials φ_ℓ and concentrations c_ℓ. In the solid phase, the variables are split into cathode potentials φ_{sc} and concentrations c_{sc} and anode potentials φ_{sa} and concentrations c_{sa}. To shorten the notation, we combine the system variables in an unknown vector $u := (\varphi_{sa}, \varphi_{sc}, \varphi_\ell, c_\ell, c_{sa}, c_{sc})$. The system variables are defined as time and space-dependent $u(\mathbf{x}, t)$ at times $t \in [0, T]$, $T \in \mathbb{R}$ and at space points $x \in \mathbb{R}$ and $(x, r) \in \mathbb{R}^2$, respectively.

Since diffusivity and conductivity must be effective values, they are modelled by taking the porosity into account:

$$\sigma_s := \hat{\sigma}_s \varepsilon_s^{\text{brugg}_s}, \quad \kappa_\ell := \hat{\kappa}_\ell \varepsilon_\ell^{\text{brugg}_\ell}, \quad D_s := \hat{D}_s \varepsilon_s^{\text{brugg}_s}, \quad D_\ell := \hat{D}_\ell \varepsilon_\ell^{\text{brugg}_\ell}. \quad (6.14)$$

The Butler-Volmer expression (6.15) couples the system Eq. (6.13).

$$ll \, j_{BV}^* = \begin{cases} zFk\left(\exp\left(\frac{\alpha z F \eta}{R\vartheta}\right) - \exp\left(\frac{-(1-\alpha)z F \eta}{R\vartheta}\right)\right) + C_{dl}\frac{\partial(\varphi_s - \varphi_\ell)}{\partial t} & \text{in } Q' \\ 0 & \text{else} \end{cases} \quad (6.15)$$

$$\eta = \varphi_s - \varphi_\ell - U_{\text{OCV}}(c_s)$$

Homogeneous Neumann conditions are applied at the boundaries, except for the outer boundaries of potentials φ_s and concentrations c_s in the solid phase:

$$\begin{aligned} \varphi_s &= 0 & \text{on } \Sigma_a := \Gamma_a \times [0, T] \\ -\sigma_s \nabla \varphi_s &= -i\,(t) & \text{on } \Sigma_c := \Gamma_c \times [0, T] \\ -D_s \frac{\partial c_s}{\partial r} &= \frac{1}{zF} j_{BV}^* & \text{on } \Sigma_{Ro} := \Gamma_{Ro,a} \cup \Gamma_{Ro,c} \times [0, T]. \end{aligned} \quad (6.16)$$

In addition, the concentrations are restricted by the initial conditions: $c_x\,(t = 0) = c_{x,0}$ in phase x in Ω and Λ, respectively. The potentials are consistently initialized at rest by the condition $j_{BV}^*\,(x, 0) = 0$. This system of four non-linearly coupled partial differential equations is solved by applying the Finite Element Method with linear test functions for spatial discretization and Backwards Euler Method for time integration. The non-linearity is solved by a damped Newton Method.

6.4 Sensitivity of Parameters

Due to the number of parameters in the model (up to 50) estimating all parameters at once is considered infeasible. It is thus necessary to simplify the model, as pointed out in Sect. 6.2. As a form of *model reduction*, we determine a reduced set of scalar parameters that still characterize the model response the most. One such tool for ranking the parameters according to their influence is "Morris-one-at-a-time" (MOAT) [4] parameter screening. This screening method yields a qualitative ranking of the parameters in terms of their global effects on the model response and enables decisions about which parameters to fix for parameter estimation. The method is based on elementary effects

$$d_i(\boldsymbol{\theta}) = \left[f(\boldsymbol{\theta}_1, \boldsymbol{\theta}_2, \ldots, \boldsymbol{\theta}_{i-1}, \boldsymbol{\theta}_i + \Delta, \boldsymbol{\theta}_{i+1}, \ldots, \boldsymbol{\theta}_k) - f(\boldsymbol{\theta}) \right] / \Delta \qquad (6.17)$$

of model responses $f(\boldsymbol{\theta})$ for a given parameter set $\boldsymbol{\theta}$ sampled along random trajectories in design space Θ. The distribution of effects associated with the ith input parameter is denoted by F_i. From this distribution, we can deduce qualitative information regarding "overall" influence—by large mean values μ_i of F_i (d_i)—and high dependency on the input—by large spread σ_i of F_i–, i.e. high interaction between parameters or high non-linearity, respectively. Since the monotonicity of the model can not be assumed, it is important to incorporate the absolute value of the measure μ^* of F_i ($|d_i|$) instead of μ—otherwise, elementary effects could cancel each other out, see [7].

For the design, the deviation in output voltage of a hybrid module driving the "New European Driving Cycle" was considered. Since input factors for the screening need not necessarily resemble parameters of the model, it was possible to combine dependent parameters pairwise, i.e. the electrolyte volume fractions ε_ℓ and solid volume fractions ε_s in anode and cathode where merged into two factors representing the sum of fractions $\varepsilon_{\ell+s}$ and their ratio $\varepsilon_{\ell:s}$, such that they may not exceed their physical limits when using simple boxing bounds.

Figure 6.2 shows the qualitative result of the screening experiment for several parameters of interest. It is thus reasonable to assume that parameter changes for many parameters will show minimal effects on the output, e.g. σ_s. Because of the high non-linearity and interaction between parameters indicated by high mean and high spread, k and D_s will most probably have a significant impact on the output.

Fig. 6.2 Influence of model parameters according to "Morris-one-at-a-time" global sensitivity analysis. Parameters to the right show more direct influence than those on the left. Parameters at the top show higher non-linear or coupled effects on the output than those near the bottom

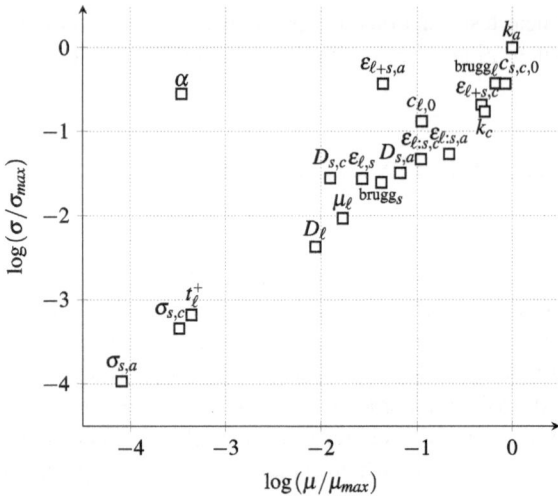

Table 6.1 Input program applied: A discharge pulse from 55 % state of charge (SoC) to ≈ 45 %

Step	Description
0	Initialize the cell to 55 % SoC
1	Discharge at $5A/Ah$ constant current for 72 s to ≈ 45 % SoC
2	Rest at zero current for 600 s,

6.5 Statistical Inversion Using MCMC

In this section, we consider the statistical inversion as described in Sect. 6.2.4 of estimating the parameters of the model defined in Sect. 6.3 matching the simulated cell voltage $f(\boldsymbol{\theta}) := \varphi_s|_{\Gamma_c}(\boldsymbol{\theta})$ to a predefined profile $\mathbf{y}$, e.g. measurements.

6.5.1 Data and Priors

For the sake of brevity, we assume the static influence, e.g. open-circuit voltage (OCV) and initial concentrations, to be completely separable from dynamic influence, e.g. diffusion and kinetic rates. This is confirmed by Speltino et al. in [9]. They describe identifying parameters of a single particle model of battery dynamics in two steps. In the first step, the equilibrium potential function of the cathode is identified from OCV measurements, assuming an equilibrium potential function of the anode from the relevant literature. The second step involves performing dynamic tests to estimate the remaining model parameters.

For estimating the data, we define a minimalistic input program $i(t)$ that we apply to both measurement and simulation, as shown in Table 6.1.

Table 6.2 Reduced parameter set under test: The table shows initial values θ_0, lower bounds θ_a, upper bounds θ_b, the applied scaling and the target values of the parameters

Name	Initial value θ_0	Lower bounds θ_a	Upper bounds θ_b	Scaling $\tilde{\theta}(\theta)$	Target $\bar{\theta}$
$D_{s,c}$	4.0e-16	1.0e-18	1.0e-13	$\log_{10}(\theta/1.0e\text{-}18)$	4.0e-17
$D_{s,a}$	4.5e-14	1.0e-15	1.0e-09	$\log_{10}(\theta/1.0e\text{-}15)$	4.5e-13
D_ℓ	6.0e-10	1e-15	1e-08	$\log_{10}(\theta/1.0e\text{-}15)$	4.0e-11
μ_ℓ	1.0e-05	1e-12	1.	$\log_{10}(\theta)$	0.1
k_c	9.0e-05	1.0e-09	1.0e-02	$\log_{10}(\theta/1.0e\text{-}09)$	9.0e-04
k_a/k_c	5.0	1.0e-02	1.0e+02	$\log_{10}(\theta/1.0e\text{-}02)$	9.0e-04/ 9.0e-04

We arrive at measurements $\mathbf{y} = \{y_i\}$ at time points t_i, which are related to the change of the cell voltage. We introduce the reduced parameter vector $\theta \in \Theta := \{\theta \in \mathbb{R}^m | \theta_a \le \theta \le \theta_b\}$. For a parameter set θ, simulation yields $f(t_j; \theta)$ at time points $t_j \le T$, which are controlled by an adaptive time stepping algorithm, that aims to give local "close-to-linear" behaviour with respect to time for internal states.

We assume that the observed voltages do not coincide exactly with the true ones of the battery at measured points, but rather that they are subject to Gaussian noise with standard deviation σ. Therefore, the observation model as in Sect. 6.2.4.

Since we do not know much about the parameter set, we choose the parameters in θ, as depicted in Table 6.2. Since the magnitude of the parameters spans a wide range, we introduce $\tilde{\theta}$ as the transformed parameter vector with logarithmic scaling applied. Since early tests revealed high correlation between k_a and k_c, we introduced a combined factor of $k_{a:c} := k_a/k_c$ instead of k_a. For the transformed parameters, we assume a *flat* prior. For the update, we use a random walk proposal kernel using Gaussian distribution initialized to $R_0 = 0.001\tilde{\theta}_0$. The measurement noise deviation and noise in (6.10) were set at $\sigma = 10^{-3}$, which is considered a realistic value.

6.5.2 Posterior Sampling

Due to the high stiffness of the model, the first run of a sampling chain is performed for parameters θ_0 using adaptive time step sizes. The resulting time points t_j are saved for later reuse in subsequent model evaluations.

The sampling is performed similar to the way proposed by Solonen et al. in [8]. It is done in parallel independent chains using Metropolis with adaptive proposal distributions $q(\hat{\theta}|\theta_n) \sim \mathcal{N}(\theta_n, R_n^2)$. The proposal deviation is set $R_n = R_0$ for $n < 20$ and for $n \ge 20$, it is updated to $R_n = chol(Cov(\theta_{start}, \ldots, \theta_{chains_n})) + \varepsilon_\Sigma$. The computation of the Covariance across all available samples starting from some index $start$ is performed in a special server, which is the only connection between the chains. To increase adaptivity, $start$ grows from 0 by 0.49 per new parameter set per chain. To ensure that R_n does not vanish, we add $\varepsilon_\Sigma = 0.001 R_0$. Because of the "flat" prior and symmetric proposals, the acceptance α becomes (6.12).

To further speed up computation, the idea of "Early Rejection" was adopted, which was first applied in [1] and [6]. First, a random number $e \in (0, 1)$ is chosen. Simulation is then executed stepwise, and in every time step t_j

$$e < \prod_{k=1}^{j} \frac{L(y_k | f(t_k; \boldsymbol{\theta}_{k+1}))}{L(y_k | f(t_k; \boldsymbol{\theta}_k))} \qquad (6.18)$$

is evaluated, and the simulation is aborted as soon as the condition is violated.

As target value $\mathbf{y}$, we chose a *simulated measurement* by running the model at the target parameters $\bar{\boldsymbol{\theta}}$. To avoid *"inverse crime"*, the reference model $\bar{f}(\bar{\boldsymbol{\theta}})$ was evaluated at different points in time $t_i \neq t_j$ than the model $f(\boldsymbol{\theta})$.

Three tests were performed to evaluate the sampling algorithm:

- *Parallel simulation without noise*—The simulation was performed in six parallel chains. The start parameters were set to $\boldsymbol{\theta}_0$. All chains were stopped at the same time after they had reached a little more than 20,000 evaluations. The first 10,000 samples per chain were discarded as a "burn-in" period to allow π (i.e. the distribution of the Markov chain) to reach equilibrium distribution.
- *Parallel simulation with noise*—The start parameters were set to $\bar{\boldsymbol{\theta}}$. The simulation was performed in five parallel chains. The simulation was aborted after all chains had reached a little more than 2,500 evaluations. The first 1,000 samples per chain were discarded as a "burn-in" period.
- *Individual simulation with noise*—Although settings were the same as for the parallel simulation with noise, this simulation was performed in three chains with no connection.

Figure 6.3 shows the results of the input profile $i(t)$ as defined in Table 6.1 and the results for the target, as well as initial and best evaluations of the simulation without noise.

6.5.3 Posterior Variability of Parameters

The statistics of the estimated parameters and uncertainties are displayed in Table 6.3. Figure 6.4 shows the posterior distributions of the parameters. Although the noiseless test was expected to show very little standard deviation, it is remarkable that the error made by interpolation of the values at time points from t_j to t_i seems to dominate the results.

Furthermore, it is evident, that the individual chains do not to sample from the entire distribution, as the statistics show very little deviation from the starting point. This can also be seen in the scatter plots in Fig. 6.5. Samples of the parallel chains cover large areas in the plots, whereas samples of the individual chain cover a slightly smaller part only and in a more dense fashion.

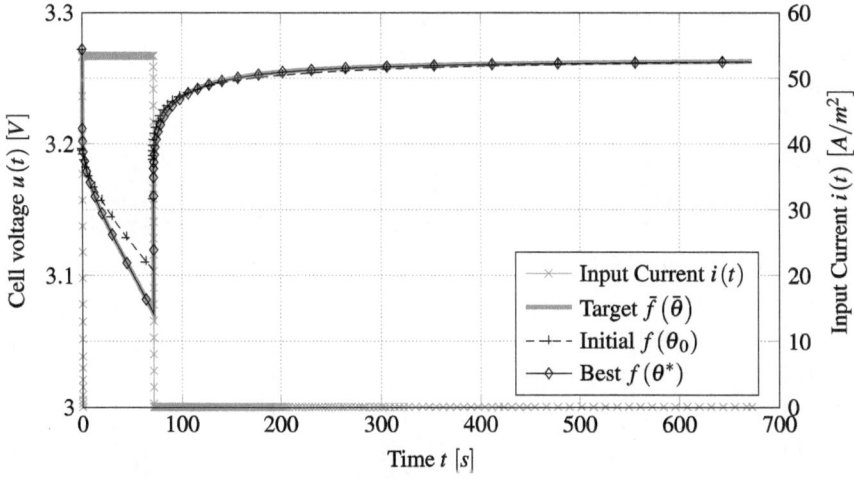

Fig. 6.3 Comparison of the initial voltage curve, synthetic target voltages and final voltages after estimation of the parameter set of interest θ

Table 6.3 Results of the uncertainties: Abbreviations are: *#A* Parallel chains without noise, *#B* Parallel chains with noise, *#C,#D,#E* individual chains with noise, *Ref* Target parameters $\tilde{\theta}$

MCMC	$\tilde{D}_{s,c}$	$\tilde{D}_{s,a}$	$\tilde{D}_\ell$	$\tilde{\mu}_\ell$	$\tilde{k}_c$	$\tilde{k}_{a:c}$
#A	1.601±0.002	2.662±0.019	4.604±0.016	−1.025±0.063	5.957±0.075	1.994±0.078
#B,	1.603±0.002	2.666±0.018	4.592±0.015	−0.969±0.056	6.044±0.093.	1.908±0.096
#C	1.603±0.002	2.662±0.016	4.603±0.005	−1.003±0.003	5.960±0.008.	1.996±0.008
#D	1.603±0.002	2.665±0.017	4.590±0.015	−0.962±0.058	6.038±0.08.	1.915±0.082
#E	1.605±0.002	2.648±0.011	4.598±0.015	−1.001±0.057	6.025±0.09	1.928±0.093
Ref	1.602±0	2.653±0	4.602±0	−1.000±0	5.954±0	2.000±0

Figure 6.5 also reveals the strong dependence between D_ℓ and μ_ℓ, and k_c and k_a, respectively. These strong correlations and the shape of the posterior distributions also indicate the logarithmic scaling and selection of $k_{a:c}$ instead of k_a. Direct sampling of all factors without the transformation and combination applied would have lead to statistical and computational inefficiency.

6.5.4 Statistical Efficiency

To assess the statistical efficiency, we use the measure of *integrated autocorrelation time* (IACT). The IACT gives the number of updates of the MCMC algorithm to give one effective independent sample. It was estimated for the posterior distributions by estimating the autocorrelation functions (ACF) for all parameters. Ideally, the ACFs for a stationary time series with little or no serial dependence reach zero quickly

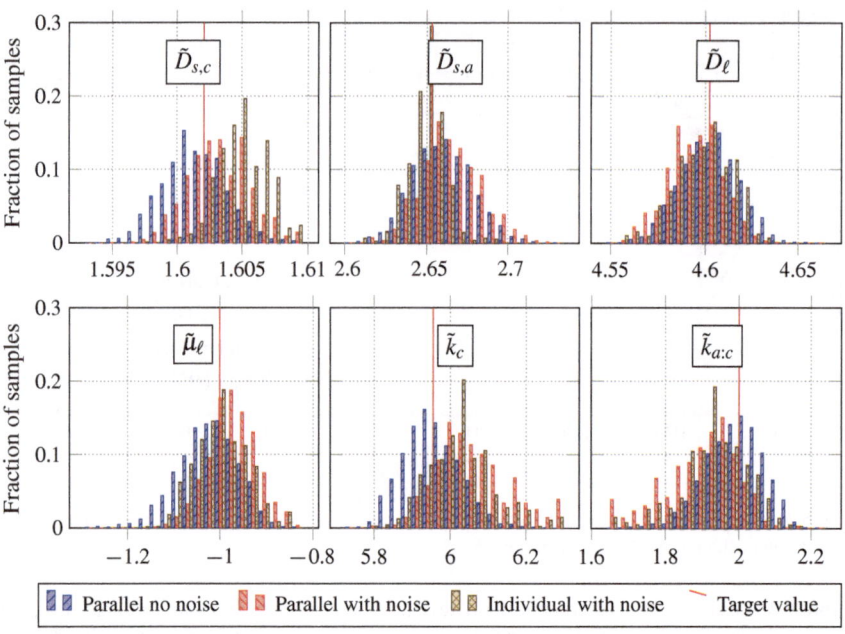

Fig. 6.4 Results of model posterior

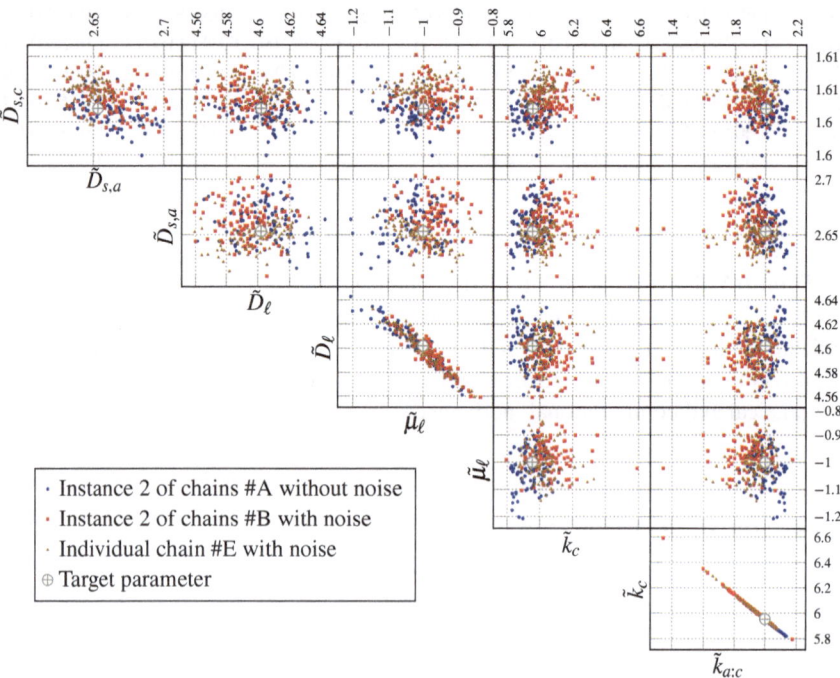

Fig. 6.5 2D scatter plots showing pairwise correlation between the parameters in θ

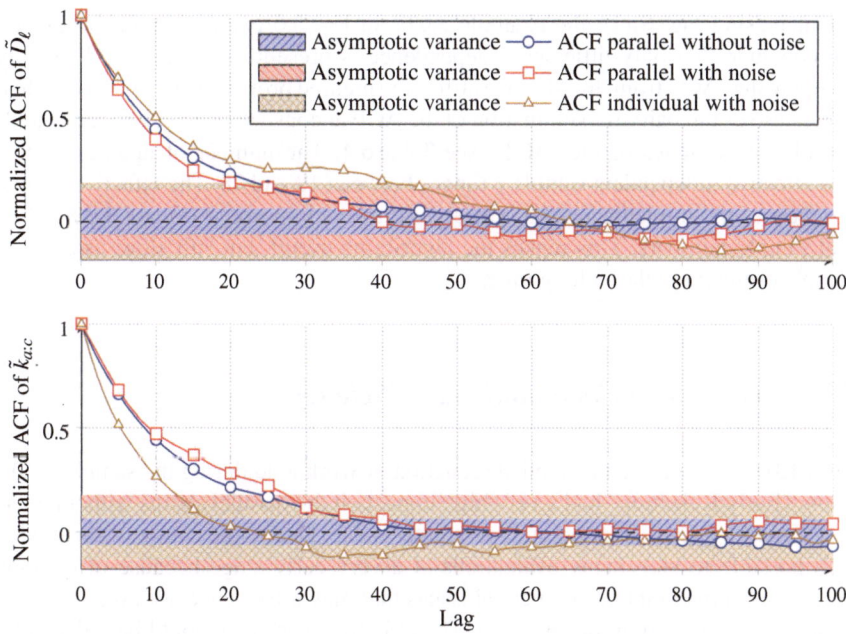

Fig. 6.6 Estimated Autocorrelation Function (ACF) for the MCMC simulations (The 95 % confidence intervals from the estimated asymptotic variance are superimposed with shaded patterns)

Table 6.4 Results of the integrated autocorrelation times τ

MCMC	$\tau - \tilde{D}_{s,c}$	$\tau - \tilde{D}_{s,a}$	$\tau - \tilde{D}_\ell$	$\tau - \tilde{\mu}_\ell$	$\tau - \tilde{K}_c$	$\tau - \tilde{K}_{a:c}$
#A-1	51.3	56.6	46.7	45.5	55.5	55.7
#A-2	67.1	49.4	52.7	55.3	48.6	48.8
#A-3	65.8	49.4	54.2	56.9	45.7	46.3
#A-4	67.5	44.6	53.7	48.2	55.6	56
#A-5	77.8	53.5	63.8	63	62.3	62.7
#A-6	45.1	45.4	60.4	60.3	53.6	53.8
#B-1	30.9	39.4	49.5	50	33.5	33.3
#B-2	35.4	41.8	45.2	54	55	54.6
#B-3	40	37.1	43.2	43.5	44.3	44.4
#B-4	40.7	28.4	41.2	40	31.9	32.2
#B-5	41.8	35.3	47.7	53.8	36.4	35.5
#C	26.1	21.3	28.6	25.5	19.5	18.9
#D	26.8	31.9	39.2	45.3	33.7	34.6
#E	30.5	54.4	49.2	53.5	26.1	26.3

for increasing lag. Since the example ACFs in Fig. 6.6 exhibit similar behaviour, we conclude that only little serial dependence is present in the chains. To finally compute the IACT from the ACF, we utilize the method proposed by [3]. This method makes use of the pairwise summation of the ACF at consecutive lags and thus yields useful overestimates of the IACT—see Table 6.4. The number of updates to give one effective independent sample τ ranges between 19 and 78. The effective sample numbers of the implemented MCMC algorithms vary between 1.3 and 5.0 % of the real sample numbers. This makes the algorithms useful, but still slow, as a single sample evaluation takes a long time.

6.5.5 Remarks on Computational Efficiency

In (6.18), we described the applied reduction of work load during the sample evaluation. "Early Rejection" had a very high impact on the parallel chains without noise that started from a distant point in parameter space. Due to the nature of the model, it is possible to stop many evaluations at a very early stage because of extreme deviations in the output or numerical issues that may arise, since not every possible parameter combination makes sense. The effective work load could be reduced by 67.7 %. That is, by using "Early Rejection", three times as many samples could be evaluated than using the regular approach.

6.6 Discussion and Conclusion

This chapter shows the applicability of parameter estimation and uncertainty quantification of lithium-ion cells by Bayesian model inversion using the Markov Chain Monte Carlo sampling approach.

We started with an introduction to parameter estimation in general, and then focussed on estimating dynamic parameters and their uncertainties in a computational model of a lithium-ion cell. We gave some insight into the modelling of the prior and the set-up of the algorithm. Due to the complexity of the model, we parallelized the approach and implemented "Early Stopping" as an additional means of reducing computing times. We compared the results of synthetic measurements and presented the statistical efficiency by investigating the integrated autocorrelation time.

The analysis of statistics in Table 6.3 and IACT in Table 6.4 indicate a sharp distribution and a higher statistical efficiency for individual chains. Only the scatter plots shown in Fig. 6.5 reveal the inferior sample coverage of the posterior in the individual chain case. This necessitates the use of parallel chains.

The proposed approach has been shown to be appropriate for investigating the dynamic properties of lithium-ion cells in the presence of noise. However, additional work must be done to incorporate stationary and quasi-stationary effects and influences, such as the open-circuit voltage and geometric quantities.

Acknowledgments The authors would like to acknowledge the financial support of the "COMET K2—Competence Centres for Excellent Technologies Programme" of the Austrian Federal Ministry for Transport, Innovation and Technology (BMVIT), the Austrian Federal Ministry of Economy, Family and Youth (BMWFJ), the Austrian Research Promotion Agency (FFG), the Province of Styria and the Styrian Business Promotion Agency (SFG).

References

1. Beskos A, Papaspiliopoulos O, Roberts GO (2006) Retrospective exact simulation of diffusion sample paths with applications. Bernoulli 12(6):1077–1098
2. Doyle M, Fuller TF, Newman J (1993) Modeling of galvanostatic charge and discharge of the lithium/polymer/insertion cell. J Electrochem Soc 140(6):1526–1533
3. Geyer CJ (1992) Practical markov chain monte carlo. Stat Sci 7(4):473–483
4. Morris M (1991) Factorial sampling plans for preliminary computational experiments. Technometrics 33(2):161–174
5. Newman J, Thomas-Alyea KE (2004) Electrochemical systems, 3rd edn., John Wiley & Sons, New York. ISBN: 978-0-471-47756-3, http://eu.wiley.com/WileyCDA/WileyTitle/productCd-0471477567.html
6. Papaspiliopoulos O, Roberts GO (2008) Retrospective markov chain monte carlo methods for dirichlet process hierarchical models. Biometrika 95(1):169–186
7. Saltelli A, Tarantola S, Campolongo F, Ratto M (2004) Sensitivity analysis in practice: a guide to assessing scientific models. Halsted Press, USA
8. Solonen A, Ollinaho P, Laine M, Haario H, Tamminen J, Järvinen H (2012) Efficient mcmc for climate model parameter estimation: Parallel adaptive chains and early rejection. Bayesian Anal 7(3):715–736
9. Speltino C, Domenico DD, Fiengo G, Stefanopoulou A (2009) Experimental identification and validation of an electrochemical model of a lithium-ion battery. In: American control conference

Chapter 7
Data-Driven Methodologies for Battery State-of-Charge Observer Design

Christoph Hametner and Stefan Jakubek

Abstract This chapter presents a data-based approach to nonlinear observer design for battery state of charge (SoC) estimation. The SoC observer is based on a purely data-driven model in order to allow for the application of the proposed concepts to any type of battery chemistry, especially when conventional physical modelling is not easily possible. In order to cope with the complex nonlinear dynamics of the battery, an integrated workflow for experiment design, model creation and automated observer design is proposed. The nonlinear battery model is constructed using a proven training algorithm based on the architecture of local model networks (LMNs). One important advantage of LMNs is that they offer local interpretability, which enables the extraction of local linear battery impedance models for automated nonlinear observer design. The proposed concepts are validated experimentally using real measurement data from a lithium-ion power cell.

7.1 Introduction

In the automotive industry, data-driven methodologies are becoming more and more important due to the constantly increasing demands. Such methods create models based on measured input and output data from the process and require little or no physical or formal information, see e.g. [29]. Especially in engine calibration, data-based approaches have been established as an important tool for systematically dealing

C. Hametner (✉)
Christian Doppler Laboratory for Model Based Calibration Methodologies, Vienna University of Technology, Wiedner Hauptstr. 8-10, 1040 Vienna, Austria
e-mail: christoph.hametner@tuwien.ac.at

S. Jakubek
Institute of Mechanics and Mechatronics, Vienna University of Technology, Wiedner Hauptstr. 8-10, 1040 Vienna, Austria
e-mail: stefan.jakubek@tuwien.ac.at

A. Thaler and D. Watzenig (eds.), *Automotive Battery Technology*,
Automotive Engineering: Simulation and Validation Methods,
DOI: 10.1007/978-3-319-02523-0_7, © The Author(s) 2014

with the growing complexity of automotive systems, see e.g. [10, 12]. In this context, the optimisation of combustion engines and hybrid electrical vehicles comprises the *calibration* of various controller parameters for both feedforward and feedback controllers in engine and hybrid control units. Thereby calibration is understood as the optimisation of vehicles and their subsystems through proper parametrisation of various controller parameters.

One important requirement for an integrated methodology for the complete calibration workflow is that both the experiment on the testbed and the model architecture must be designed such that the model is able to cover all relevant effects, and all parameter interactions can be taken into account in the optimisation procedure. Such a *model-based calibration workflow* consists of the following steps: Experiment design, nonlinear system identification and controller/observer analysis and design.

Besides engine calibration, the optimisation of hybrid components has become an important issue in recent years. Hybrid electrical vehicles require an accurate online observation of the electric power supply. In this context, the development of the battery management system (BMS) and the energy management system (EMS) is thus a challenging task. One essential part of the BMS is a *battery model*, which must be accurate under the specific loads and environmental conditions. One of the most important functions of the BMS is determining the state of charge (SoC) of the battery, as well as the charge and discharge control. Knowledge about the state of charge, which cannot be measured directly, is thus essential in order to extend battery life and preserve the usable capacity.

The present work describes the model-based calibration workflow using data-driven models in general and presents the adaptation/extension of the proposed concepts to battery modelling and the design of the associated nonlinear observer (see also [11]). The remainder of this chapter is structured as follows: Sect. 7.2 provides an overview on the three major steps of the data-driven calibration workflow. Section 7.3 presents the application of the proposed concepts for SoC observer design and demonstrates the performance using real measurement data from a lithium-ion cell.

7.2 Data-Driven Calibration Workflow

Since conventional physical modelling is difficult in many situations, black-box and grey-box-oriented nonlinear system identification procedures have emerged as a feasible alternative in model-based calibration. One important advantage of *grey-box* approaches is that they enable reduced model complexity, if physical (or other) insight into the nature of the object is available, [19]. In this context, local model networks (LMNs) have proven to be a powerful tool, e.g. [4, 8, 17, 22]. LMNs can adapt to the complexity of the problem in a highly efficient way, and then also make it easy to incorporate prior (physical) knowledge.

However, significant expert knowledge and experience is required to use identification algorithms, even though these methods have been well developed. In order to facilitate the operation of system identification methods for non-experts, a *unified and generic* model-based calibration workflow is proposed here. The major steps for such an integrated methodology for model-based calibration are described below:

Experiment design: The design of experiments (DoE) is an important prerequisite for data-based modelling approaches and thus constitutes the first step in the model-based calibration workflow. The target of experiment design is the proper excitation of the unknown system such that the global model behaviour can be determined from measured input and output data. Due to the high dimensionality of the problems and the high costs involved with testbed time, the number of measurements must be kept to a minimum. Every single design point must be placed such that the maximum possible information is gained. The quality of the experiment design is therefore decisive for the model quality and the subsequent optimisation procedure.

Nonlinear system identification: The goal of nonlinear system identification is to obtain a process model from measured data. This can involve the *parametrisation* of a model within a given structure alone (e.g. a difference equation) or also the superordinated determination of the *structure* itself. Traditionally, mathematical models have been obtained using physical laws or other theoretical knowledge about the system (white-box models). This method requires a technical expert for modelling and simulation because of the demanding complexity of real systems. However, in many applications conventional modelling is difficult or even impossible due to the lack of precise, formal knowledge about the system. Especially for components such as combustion engines or batteries, the *rapid* development of physical models in an economically competitive environment is quite difficult. Thus, model-based calibration uses black and grey-box-oriented nonlinear system identification procedures. Such methods create models based on measured input and output data of the process and require little or no physical or formal information.

Controller and observer analysis and design: The final step in the model-based calibration workflow is the calibration of maps (feedforward control, in the most general sense) and the design/parametrisation of controllers (feedback control) and observers. One immediate advantage of using a model-based approach for observer design lies in the fact that the task can be accomplished without using the actual plant.

The following section describes the proposed concepts for model generation (i.e. optimal experiment design and nonlinear system identification) and observer design using the architecture of LMNs in detail.

7.3 State-of-Charge Observer Design

7.3.1 Experiment Design

A suitable experiment design is an important prerequisite for data-based modelling approaches. In order to achieve a model which describes the underlying process properly, the system inputs have to excite the process in such a way that all relevant dynamics and nonlinearities become visible from the measured data. At the same time, however, the experiment design must ensure compliance with the operational constraints and limitations of the still unknown system under consideration.

Basically, there are *model-free* and *model-based* concepts for the experiment design:

Model-free DoE: If no information about the underlying process is available, space filling designs are commonly used in nonlinear *static* system identification. These approaches are generally targeted to cover the whole input space uniformly. For nonlinear *dynamic* systems, amplitude modulated pseudorandom binary signals (APRBS), see e.g. [23], are typically chosen to excite the system dynamics while the desired operating range is covered.

Model-based DoE: Model-based DoE is more specifically tailored to the process to be identified in that a prior process model (or at least a model structure) is used to maximize the information gained from experiments. Thus, the signal is adjusted to a specific *prior process model* such that the model parameters can be estimated from measured data with minimal variance, see e.g. [7, 25]. In this context, the Fisher information matrix gives a statement about the information content of data with respect to the covariance of estimated parameters.

For the model-based calibration workflow, the second approach was chosen, which itself is model-based. Thus, the fundamental discrepancy is that a reference model (prior process model) of the underlying process is required for the modelling of the very same process. However, model-based experiment design offers some favorable properties which are highlighted in the following. First, when a physical model (e.g. from the design phase of a process) or a model from a similar process is available, model-based DoE helps to increase the information content of the measured data while reducing the experimentation effort. Second, the compliance with operational constraints during the experiment (e.g. in order to prevent damage to the plant) can be taken into account using a model-based approach.

For battery model identification, the whole operating range (cell current, SoC) of the battery has to be covered. One of the main challenges in this context is that the SoC excitation and the associated operating range directly depend on the excitation of the cell current itself. While APRB signals are typically used to track the nonlinearities of the process under consideration, this strategy is no longer feasible for batteries, since the dependence between cell current and SoC is not taken into account, leading to insufficient coverage of the operating range of the battery. Consequently, the experiment design has to assure that the variation of the cell current both yields a proper excitation of the cell and covers the desired SoC operating range, [11].

In [14], the topic of optimal model-based DoE for LMNs is addressed. Thereby, the optimisation of the experiment design is based on the Fisher information matrix of the LMN, and the proposed concepts are targeted to generate informative data and to reduce the experimentation effort as much as possible. For optimal experiment design for batteries, the gradient based optimisation from [14] cannot be applied directly, due to the special situation that the excitation of one input channel (state of charge) is determined by another input signal (cell current). Thus, a model-based DoE approach using a *linear* dynamic battery model, which describes the nonlinear system behaviour of the cell terminal voltage based on the charge/discharge current, was chosen. The linear battery model was identified using data from preliminary tests (step responses). Note that other battery models (e.g. simple equivalent circuit models) could also be used for the model-based DoE procedure. However, the results indicate that even a simple linear model helps to improve the model quality significantly, see Sect. 7.3.3.

Based on the linear (prior) battery model, the goal of the optimal experiment design is to increase the information content of the data while reducing the testing time is. Thus, the following performance function is minimised

$$J_{opt} = \alpha \frac{J_{FIM,init}}{J_{FIM,opt}} + (1 - \alpha) \frac{T_{opt}}{T_{init}} \tag{7.1}$$

where $J_{FIM,init}$ and $J_{FIM,opt}$ denote the determinant of the Fisher information matrix $J_{FIM} = \det(\mathscr{I})$ of the initial design and the optimised design, respectively. Thereby, the FIM is based on the partial derivative of the model output with respect to the model parameters, c.f. [7]:

$$\mathscr{I} = \frac{1}{\sigma^2} \sum_{k=1}^{N} \frac{\partial \hat{y}(k,\theta)}{\partial \theta} \frac{\partial \hat{y}(k,\theta)}{\partial \theta}^T . \tag{7.2}$$

In (7.1) the duration of the testbed run is taken into account by T_{init} and T_{opt}. The design parameter $0 \le \alpha \le 1$ in (7.1) is chosen based on the tradeoff between accuracy and measurement effort: When α is increased, the information content is increased; for $\alpha - 0$ only the testing time is reduced.

The optimisation itself is focused on the appropriate sequential arrangement of predefined SoC and current levels using the simulated annealing method. These predefined levels are chosen such that a proper distribution of the cell current excitation is obtained and the complete SoC operating range is covered. The sign of the cell current (charging or discharging) in each level and the duration of the associated charge/discharge pulse are obtained from the corresponding SoC level and the previous design point (i.e. the previous SoC). Thus, the testing time depends directly on the sequential arrangement of the SoC levels and the associated current levels. Similar to [21], the search begins with an initial design and proceeds through examination of

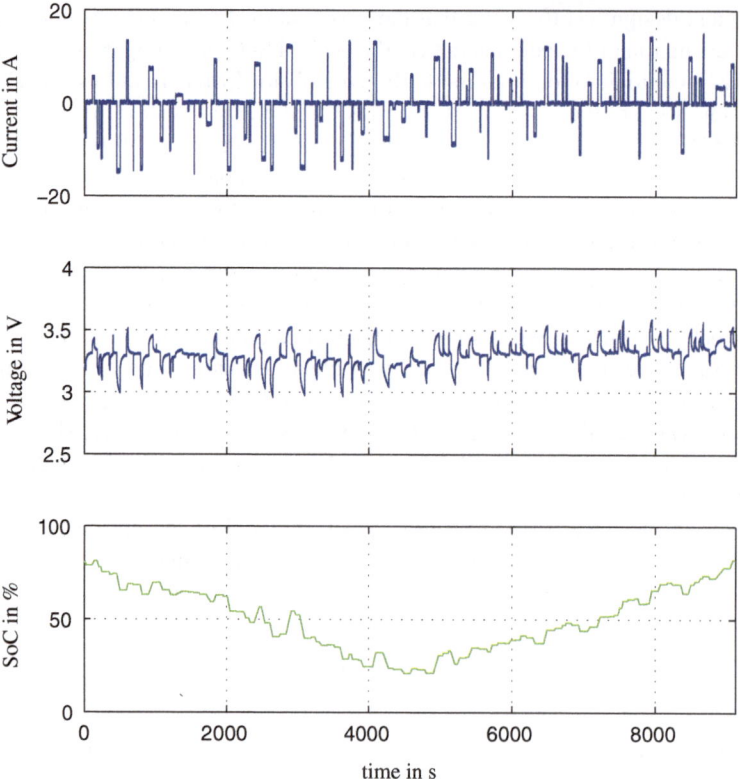

Fig. 7.1 Experiment design: Optimised training data

a sequence of designs, each generated as a perturbation of the preceding one. Thus, a new candidate state is obtained from a random exchange of the current and SoC levels. The probability of making a transition from the current state to the new state is specified by an acceptance probability function.

Figure 7.1 depicts the optimised training data record (i.e. the choice of the excitation signals and the measured battery voltage). Compared to the initial design, the determinant of the Fisher information matrix was increased by a factor of five, while the testing time was reduced by about 7 %.

7.3.2 Data-Driven Battery Modelling

The second step in the model-based calibration workflow is the nonlinear system identification procedure itself.

Fig. 7.2 Validity functions of an LMN comprising six local models

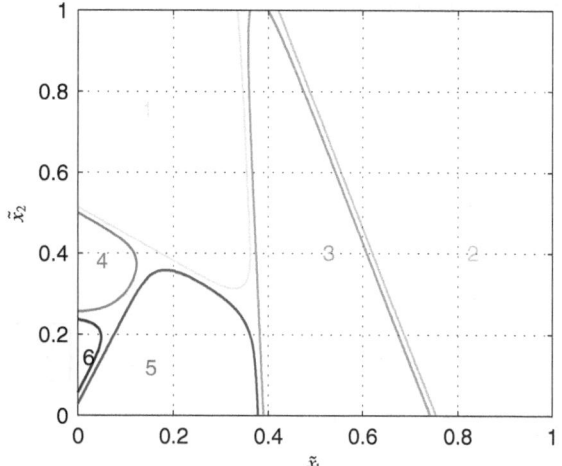

One integral part of the SoC observer is a mathematical cell model, which describes the nonlinear dynamic behaviour of the cell terminal voltage based on the charge/discharge current. Such a model allows, which enables the prediction of the nonlinear system dynamics of the traction battery, is required for SoC estimation, since it is not possible to measure the SoC directly. Typical modelling approaches include electro-chemical models (see e.g. [1, 6, 18]) and equivalent circuit models (see e.g. [16]). However, since it is not always easy to achieve time-efficient parametrisation and real-time application, a data-driven approach has been chosen, which allows for the application of the proposed concepts for any type of battery chemistry.

The dynamic identification algorithm used in this work is based on the architecture of LMNs, [9]. For practical application, both the integration of prior knowledge and the interpretability of the individual local models are of great interest. The construction of LMNs is based on partitioning the operating space into a number of operating regimes, see Fig. 7.2. The global model output is then formed by a weighted combination of local models, each of which is valid in a certain operating regime. The architecture of LMNs represents an excellent approach for the integration of various knowledge sources. Accordingly, the complexity of the identification procedure can be reduced significantly when prior knowledge about the underlying system is available. In practice, this means that some expected behaviour of the modelled system helps to choose a suitable scheduling variable (e.g. $\tilde{x}_1$ and $\tilde{x}_2$ in Fig. 7.2), select the structure and find appropriate parameters for the model.

This section briefly reviews the proposed LMN and the identification of the battery model for the purpose of SoC estimation. The nonlinear model describes the dynamic behaviour of the terminal voltage $U(t)$ based on the charge/discharge current $I(t)$ and other factors (e.g. temperature and SoC). The LMN interpolates between different local models, each of which valid in a certain region of the input space. Thus, the battery cell model is based on a partitioning into several local operating regimes (local

Fig. 7.3 Logistic discriminant tree

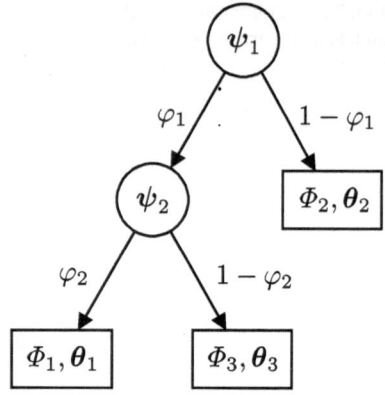

linear impedance models), represented by the dominant influence of the scheduling variables, such as SoC, temperature, etc. This strategy makes it possible to capture the highly nonlinear dynamic complexity in a computationally efficient way.

In general, each local model of the LMN—indicated by subscript i—consists of two parts: The *validity function* $\Phi_i(\tilde{\mathbf{x}}(k))$ and its *model parameter vector* θ_i. Thereby, Φ_i defines the region of validity of the i-th local model.

The *local* estimate for the output is obtained by

$$\hat{y}_i(k) = \mathbf{x}^T(k)\theta_i, \tag{7.3}$$

where $\mathbf{x}^T(k)$ denotes the regressor vector. In dynamic system identification, the regressor vector $\mathbf{x}(k)$ comprises past system inputs and outputs.

All local estimations $\hat{y}_i(k)$ are used to form the global model output $\hat{y}(k)$ by weighted aggregation

$$\hat{y}(k) = \sum_{i=1}^{M} \Phi_i(k)\hat{y}_i(k), \tag{7.4}$$

where

$$\Phi_i(k) = \Phi_i(\tilde{\mathbf{x}}(k)) \tag{7.5}$$

and M denotes the number of local linear models. Thereby, the elements in $\tilde{\mathbf{x}}(k)$ span the so-called partition space and are chosen on the basis of prior knowledge about the process and the expected structure of its nonlinearities.

The computation of the validity functions Φ_i is based on a logistic discriminant tree. Figure 7.3 depicts a model tree with three local models. Each node corresponds to a split of the partition space into two parts, and the free ends of the branches represent the actual local models with their parameter vector θ_i and their validity functions Φ_i. The overall nonlinear model thus comprises M local models and $M-1$ nodes that determine their regions of validity.

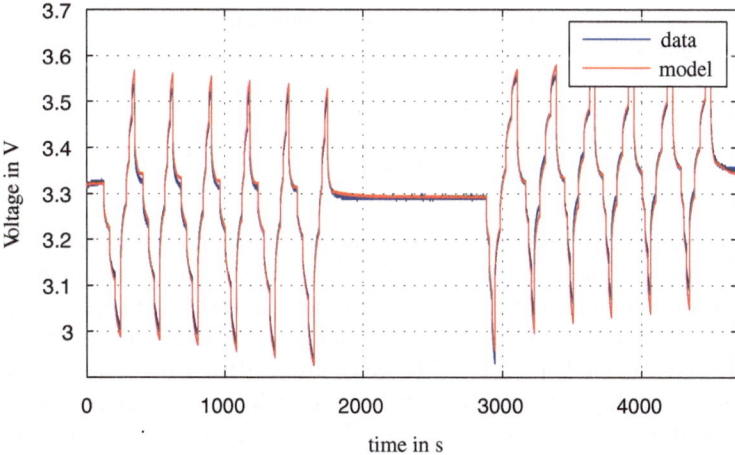

Fig. 7.4 Model validation/generalisation: Comparison of simulated model output and measured terminal voltage

$$\Phi_1 = \varphi_1\varphi_2, \tag{7.6}$$

$$\Phi_2 = 1 - \varphi_1, \tag{7.7}$$

$$\Phi_3 = \varphi_1(1 - \varphi_2). \tag{7.8}$$

For the representation of the discriminant function in the d-th node, a logistic sigmoid activation function is chosen, c.f. [3]:

$$\varphi_d(\tilde{\mathbf{x}}(k)) = \frac{1}{1 + \exp(-a_d(\tilde{\mathbf{x}}(k)))} \tag{7.9}$$

with

$$a_d(\tilde{\mathbf{x}}(k)) = \begin{bmatrix} 1 & \tilde{\mathbf{x}}^T(k) \end{bmatrix} \begin{bmatrix} \psi_{d0} \\ \tilde{\psi}_d \end{bmatrix}. \tag{7.10}$$

Here, $\tilde{\psi}_d^T = \begin{bmatrix} \psi_{d1} \dots \psi_{dp} \end{bmatrix}$ denotes the weight vector, and ψ_{d0} is called the bias term. The discriminant functions φ_d are used to calculate the validity functions Φ_i, c.f. [26]. The validity functions for the layout in Fig. 7.3 are obtained by (7.6), (7.7) and (7.8).

The training (i.e. the parametrisation) of the battery model (the LMN) is then based on a nonlinear optimisation algorithm (for a more detailed description, please refer to [9] and [10]). The performance of the LMN is highlighted using real measurement data from the lithium-ion cell in Fig. 7.4, which depicts a comparison of measured and simulated model output (validation data). The results indicate the excellent generalisation capabilities of the proposed LMN training algorithm.

7.3.3 Nonlinear Observer Design

LMNs provide a basis for the development of systematic approaches to stability analysis and controller/observer design in view of powerful conventional control theory. Thus, one immediate advantage of using a model-based approach for controller and observer design lies in the fact that the task can be accomplished without using the actual plant. In [13], LMN-based controller design and associated stability analysis methodologies for LMNs, which are typically based on an associated *local* controller for every local model, are described. In combination with the proposed LMN, different *observer* structures can be used, whereby the architecture and interpretability of the local models as a local linearisation of the process helps to reduce the model/observer complexity. These observer structures include the extended Kalman filter (EKF) and the fuzzy observer:

- One widely used approach for SoC estimation of batteries is the *extended Kalman filter* in combination with equivalent circuit models, see e.g. [2, 15, 31, 32]. The idea of the EKF is to apply conventional Kalman filtering to a nonlinear system whereby the filter gain is computed using the local Jacobian of the nonlinear model.
- Similar to the design of local controllers, a *fuzzy observer* can be used for state estimation with LMNs, see e.g. [5, 24, 27, 28]. A local observer is designed for each local linear model using standard Kalman filter theory. The global filter is then derived from a linear combination of the local filters, [28]. Thus, compared to the EKF, the *local* observers are time-invariant, which greatly reduces the computational complexity of the global filter. Another important advantage of the fuzzy observer architecture is that the stability analysis of the nonlinear observer is possible based on Lyapunov stability theory, see e.g. [20, 30].

This section describes the design of a fuzzy observer for SoC estimation. The SoC observer is based on a combination of the terminal voltage model (as described in Sect. 7.3.2) and a relative SoC model

$$SoC(t) = SoC_0 + \int_{\tau=0}^{t} \frac{\eta_I(I)I(\tau)}{C_n} d\tau \qquad (7.11)$$

where SoC_0 denotes the initial SoC, $I(t)$ is the instantaneous cell current, C_n is the nominal cell capacity, and $\eta_I(I)$ is the coulombic efficiency (see also [11]). A schematic of this approach is given in Fig. 7.5, where T_{Bat} represents the measured temperature, and U_{Bat} and $\hat{U}_{Bat}$ define the measured and predicted (simulated) battery voltage, respectively. Thus, the SoC correction is obtained from a comparison of the *actual* terminal voltage to the output of the model.

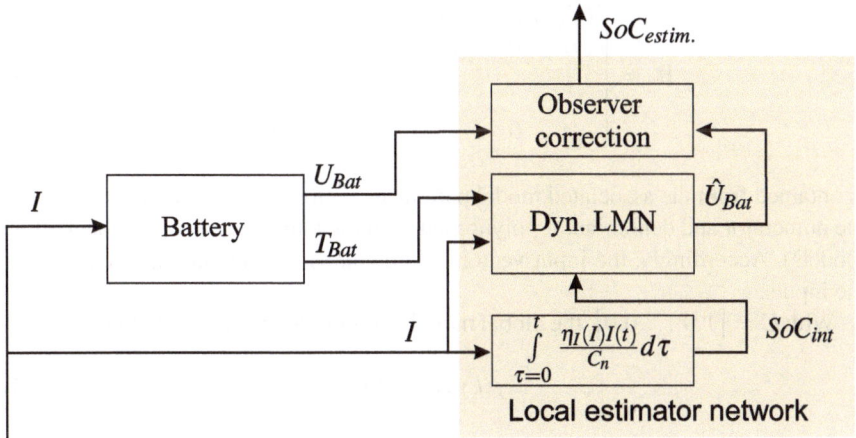

Fig. 7.5 Schematic of SoC observer architecture

The nonlinear observer design involves representation of the nonlinear system as a local linear state space model, [27]. In combination with the (relative) SoC model (7.11), an augmented state-space formulation of the nonlinear model is obtained. Using the relation $SOC(k) = SOC(k-1) + \frac{T_s}{C_n}i(k)$, the augmented state vector (7.12) also comprises SoC, which was originally used as a model input for the training of the LMN. In addition, the state vector comprises past elements of the system output:

$$\mathbf{z}(k-1) = \begin{bmatrix} y(k-1) \\ y(k-2) \\ \vdots \\ y(k-n) \\ SOC(k-1) \end{bmatrix}. \tag{7.12}$$

For each local model, the system matrix

$$\mathbf{A}_i = \begin{bmatrix} a_{1,i} & a_{2,i} & \cdots & a_{n,i} & b_{SOC,i} \\ 1 & 0 & \cdots & 0 & 0 \\ 0 & 1 & \cdots & 0 & 0 \\ \vdots & \vdots & \ddots & \vdots & \vdots \\ 0 & 0 & \cdots & 0 & 1 \end{bmatrix} \tag{7.13}$$

and the input matrix

$$
\mathbf{B}_i = \begin{bmatrix} b_{10,i} & b_{11,i} & \dots & b_{20,i} & b_{21,i} & \dots & b_{2m,i} & c_i \\ 0 & 0 & \dots & 0 & 0 & \dots & 0 & 0 \\ \vdots & \vdots & \ddots & \vdots & \vdots & \ddots & \vdots & \vdots \\ \frac{T_s}{C_n} & 0 & \dots & 0 & 0 & \dots & 0 & 0 \end{bmatrix}
\tag{7.14}
$$

is obtained from the associated model parameter vector θ_i (i.e. from the elements of the numerator and denominator polynomials and the affine term c_i of the local affine models). Accordingly, the input vector $\mathbf{u}(k)$ contains current and past elements of the inputs.

With $\mathbf{C} = \begin{bmatrix} 1 & 0 & \dots & 0 & 0 \end{bmatrix}$, the global model output is defined by the output equation

$$
y(k) = \mathbf{C}\mathbf{z}(k)
\tag{7.15}
$$

and the state equation (of the local models):

$$
\mathbf{z}(k) = \sum_{i=1}^{M} \Phi_i(k-1) \{ \mathbf{A}_i \mathbf{z}(k-1) + \mathbf{B}_i \mathbf{u}(k) \}.
\tag{7.16}
$$

Based on the augmented state-space formulation of the LMN, the state estimate $\hat{\mathbf{z}}(k)$ is given by the following equations:

$$
\mathbf{z}_i^*(k) = \mathbf{A}_i \hat{\mathbf{z}}(k-1) + \mathbf{B}_i \mathbf{u}(k)
\tag{7.17}
$$

and

$$
\hat{\mathbf{z}}(k) = \sum_{i=1}^{M} \Phi_i(k-1) \{ \mathbf{z}_i^*(k) + \mathbf{K}_i[y(k) - \hat{y}(k)] \}
\tag{7.18}
$$

where

$$
\hat{y}(k) = \sum_{i=1}^{M} \Phi_i(k-1) \mathbf{C} \mathbf{z}_i^*(k)
\tag{7.19}
$$

In (7.18), the matrix $\mathbf{K}_i$ defines the Kalman filter gain of the i-th local model, which is obtained from

$$
\mathbf{K}_i = \mathbf{A}_i \mathbf{P}_i^T \mathbf{C}^T (\mathbf{C} \mathbf{P}_i^T \mathbf{C}^T + \mathbf{R}^T)^{-1}
\tag{7.20}
$$

where $\mathbf{P}_i$ is the solution of the discrete-time algebraic Riccati equation (DARE)

$$
\mathbf{A}_i \mathbf{P}_i \mathbf{A}_i^T - \mathbf{P}_i - \mathbf{A}_i \mathbf{P}_i \mathbf{C}^T (\mathbf{C} \mathbf{P}_i \mathbf{C}^T + \mathbf{R})^{-1} \mathbf{C} \mathbf{P}_i \mathbf{A}_i^T + \mathbf{Q} = 0.
\tag{7.21}
$$

with the covariance matrices of the measurement noise denoted by $\mathbf{Q}$ and $\mathbf{R}$.

The performance of the fuzzy observer for SoC estimation is again demonstrated by means of the validation data record. Now, the SoC is assumed to be unknown, and the initial state of SoC is chosen at random. The upper graph in Fig. 7.6 shows

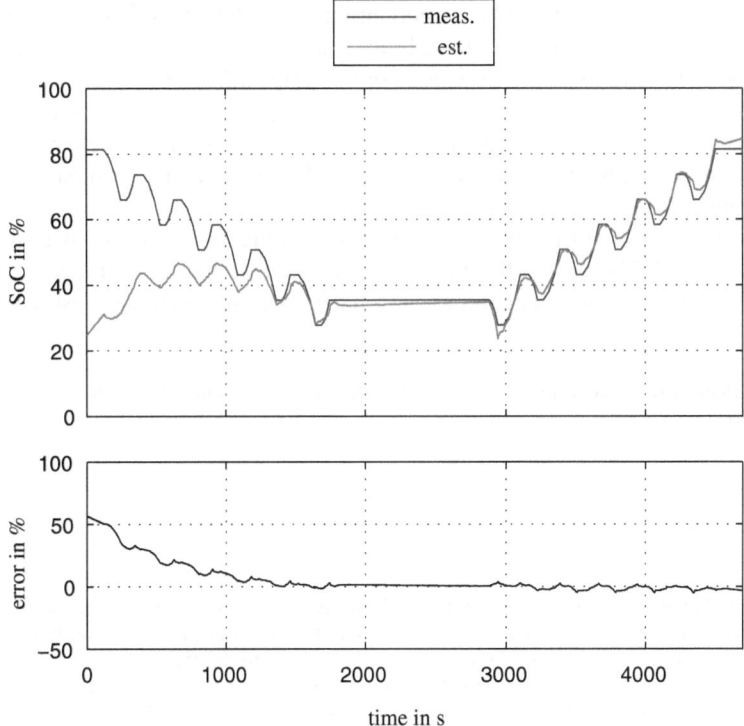

Fig. 7.6   Generalisation data: SoC estimation/correction after random initialisation

a comparison of the actual and estimated SoC of the lithium-ion cell. The estimated SoC converges to the correct value and the LMN-based fuzzy observer accurately estimates the unknown SoC.

7.4 Conclusion

This chapter addressed the topic of data-based battery modelling and the associated state of charge observer design. The complete data-driven calibration workflow, which consists of the experiment design, nonlinear system identification and observer design, was described.

As an important prerequisite for data-based modelling techniques, the target of the *experiment design* is a proper excitation of the unknown system. A model-based experiment design approach, which itself is based on a simple battery model, was chosen in order to increase the information content of the data while simultaneously reducing the testing time. Based on the measured data record, the structure and parameters of the nonlinear battery model were identified using the proposed local model network training algorithm. One important feature of LMNs is that they provide a basis for systematic approaches to stability analysis and observer design. Based on an augmented state space representation of the LMN, a fuzzy observer for SoC estimation is designed. The greatest advantage over the widely used extended Kalman filter is that the local observers are time-invariant, and linearisation at every sampling instance is not required, which greatly reduces the computational complexity of the global filter.

The performance of the proposed concepts was highlighted using measurement data from the lithium-ion cell. The results indicate that the battery model provides excellent generalisation capabilities, and fuzzy observer accurately estimates the SoC.

Acknowledgments This work was supported by the Christian Doppler Research Association and AVL List GmbH, Graz.

References

1. Arora P, Doyle M, Gozdz AS, White RE, Newman J (2000) Comparison between computer simulations and experimental data for high-rate discharges of plastic lithium-ion batteries. J Power Sources 88(2):219–231. doi:10.1016/S0378-7753(99)00527-3
2. Bhangu B, Bentley P, Stone D, Bingham C (2005) Nonlinear observers for predicting state-of-charge and state-of-health of lead-acid batteries for hybrid-electric vehicles. IEEE Trans Veh Technol 54(3):783–794. doi:10.1109/TVT.2004.842461
3. Bishop CM (1995) Neural networks for pattern recognition. Oxford University Press, USA
4. Brown M, Lightbody G, Irwin G (1997) Local model networks for nonlinear system identification. In: IEE colloquium on industrial applications of intelligent control (Digest No: 1997/144), pp 4/1–4/3. doi:10.1049/ic:19970785
5. Chen G, Xie Q, Shieh LS (1998) Fuzzy kalman filtering. Inf Sci 109(14):197–209. doi:10.1016/S0020-0255(98)10002-6
6. Gomadam PM, Weidner JW, Dougal RA, White RE (2002) Mathematical modeling of lithium-ion and nickel battery systems. J Power Sources 110(2):267–284. doi:10.1016/S0378-7753(02)00190-8
7. Goodwin G, Payne R (1977) Dynamic system identification: experiment design and data analysis. In: Mathematics in science and engineering, vol 136. Academic Press, New York
8. Gregorcic G, Lightbody G (2007) Local model network identification with gaussian processes. IEEE Trans Neural Netw 18(5):1404–1423. doi:10.1109/TNN.2007.895825
9. Hametner C, Jakubek S (2007) Neuro-fuzzy modelling using a logistic discriminant tree. In: American control conference, 2007 ACC '07, pp 864–869. doi:10.1109/ACC.2007.4283048
10. Hametner C, Jakubek S (2013) Local model network identification for online engine modelling. Inf Sci 220(0):210–225, doi:10.1016/j.ins.2011.12.034, http://www.sciencedirect.com/science/article/pii/S0020025512000138, online Fuzzy machine learning and data mining
11. Hametner C, Jakubek S (2013) State of charge estimation for lithium ion cells: design of experiments, nonlinear identification and fuzzy observer design. J Power Sources 238(0):413–421. doi:10.1016/j.jpowsour.2013.04.040
12. Hametner C, Nebel M (2012) Operating regime based dynamic engine modelling. Control Eng Pract 20(4):397–407. doi:10.1016/j.conengprac.2011.10.003, http://www.sciencedirect.com/science/article/pii/S0967066111002085
13. Hametner C, Mayr CH, Kozek M, Jakubek S (2013) PID controller design for nonlinear systems represented by discrete-time local model networks. Int J Control 86(9):1453–1466.

        doi:10.1080/00207179.2012.759663,   http://www.tandfonline.com/doi/abs/10.1080/00207
        179.2012.759663
14. Hametner C, Stadlbauer M, Deregnaucourt M, Jakubek S, Winsel T (2013) Optimal experiment
        design based on local model networks and multilayer perceptron networks. Eng Appl Artif Intell
        26(1):251–261. doi:10.1016/j.engappai.2012.05.016, http://www.sciencedirect.com/science/
        article/pii/S0952197612001224
15. Han J, Kim D, Sunwoo M (2009) State-of-charge estimation of lead-acid batteries using an
        adaptive extended kalman filter. J Power Sources 188(2):606–612. doi:10.1016/j.jpowsour.
        2008.11.143
16. Hu Y, Yurkovich BJ, Yurkovich S, Guezennec Y (2009) Electro-thermal battery modeling and
        identification for automotive applications. In: ASME conference proceedings 2009(48937), pp
        233–240. doi:10.1115/DSCC2009-2610
17. Johansen TA, Foss BA (1995) Identification of non-linear system structure and parameters
        using regime decomposition. Automatica 31(2):321–326. doi:10.1016/0005-1098(94)00096-
        2
18. Klein R, Chaturvedi N, Christensen J, Ahmed J, Findeisen R, Kojic A (2010) State estimation
        of a reduced electrochemical model of a lithium-ion battery. In: American control conference
        (ACC), 2010, pp 6618–6623
19. Ljung L (2008) Perspectives on system identification. In: Proceedings of the 17th IFAC world
        congress
20. Mayr C, Hametner C, Kozek M, Jakubek S (2011) Piecewise quadratic stability analysis for
        local model networks. In: 2011 IEEE international conference on control applications (CCA),
        pp 1418–1424. doi:10.1109/CCA.2011.6044503
21. Morris M (1995) Exploratory designs for computational experiments. J Stat Plan Infer
        43(3):381–402. doi:10.1016/0378-3758(94)00035-T
22. Murray-Smith R, Johansen TA (1997) Multiple model approaches to modelling and control.
        Taylor & Francis, London
23. Nelles O (2002) Nonlinear system identification, 1st edn. Springer, Berlin
24. Polansky M, Ardil C (2007) Robust fuzzy observer design for nonlinear systems. Int J Math
        Comput Sci 3:1
25. Pronzato L (2008) Optimal experimental design and some related control problems. Automatica
        44(2):303–325
26. Pucar P, Millnert M (1995) Smooth hinging hyperplanes—an alternative to neural networks.
        In: Proceedings of the 3rd ECC
27. Senthil R, Janarthanan K, Prakash J (2006) Nonlinear state estimation using fuzzy kalman
        filter. Ind Eng Chem Res 45(25):8678–8688. doi:10.1021/ie0601753
28. Simon D (2003) Kalman filtering for fuzzy discrete time dynamic systems. Appl Soft Comput
        3(3):191–207. doi:10.1016/S1568-4946(03)00034-6
29. Sjoberg J, Zhang Q, Ljung L, Benveniste A, Deylon B (1995) Nonlinear black-box modeling
        in system identification: a unified overview. Automatica 31:1691–1724
30. Tanaka K, Ikeda T, Wang H (1998) Fuzzy regulators and fuzzy observers: relaxed stability
        conditions and LMI-based designs. IEEE Trans Fuzzy Syst 6(2):250–265. doi:10.1109/91.
        669023
31. Vasebi A, Partovibakhsh M, Bathaee SMT (2007) A novel combined battery model for state-
        of-charge estimation in lead-acid batteries based on extended kalman filter for hybrid electric
        vehicle applications. J Power Sources hybrid Electr Veh 174(1):30–40. doi:10.1016/j.jpowsour.
        2007.04.011
32. Vasebi A, Bathaee S, Partovibakhsh M (2008) Predicting state of charge of lead-acid batteries
        for hybrid electric vehicles by extended kalman filter. Energy Convers Manage 49(1):75–82,
        doi:10.1016/j.enconman.2007.05.017

Erratum to: Application-Related Battery Modelling: From Empirical to Mechanistic Approaches

Franz Pichler and Martin Cifrain

Erratum to:
Chapter 4 in: A. Thaler and D. Watzenig (eds.), *Automotive Battery Technology*, DOI 10.1007/978-3-319-02523-0_4

The word "textlayer" in Eq. (4.23) of Chap. 4 should read as "layer":

$$Gstack = (\sum_{layer} \frac{1}{G_{layer}})^{-1}. \tag{4.23}$$

The online version of the original chapter can be found under DOI 10.1007/978-3-319-02523-0_4

F. Pichler (✉) · M. Cifrain
Virtual Vehicle Research Center, Graz, Austria
e-mail: franz.pichler@v2c2.at

M. Cifrain
e-mail: martin.cifrain@v2c2.at

A. Thaler and D. Watzenig (eds.), *Automotive Battery Technology*,   E1
Automotive Engineering: Simulation and Validation Methods,
DOI: 10.1007/978-3-319-02523-0_8, © The Author(s) 2014

About the Editors and Authors

Editors

Alexander Thaler studied electric engineering at the Graz University of Technology and the University of Leoben, where he received his PhD degree on the field of electric drive trains. In 2005 he began working for MAGNA Steyr, where he was part of the product development team for lithium-ion battery systems. From 2009 onwards he was responsible for advanced development and concept development for battery systems in automotive application. He joined the VIRTUAL VEHICLE Research Center in 2011 and is currently the team leader of the battery group, where he coordinates the research work in this field.

Daniel Watzenig is an Associate Professor at the Institute of Electrical Measurement and Measurement Signal Processing at Graz University of Technology. He is currently the divisional director and scientific head of the automotive electronics and embedded software department of the VIRTUAL VEHICLE Research Center in Graz. His research interests are focussed on noninvasive measurement techniques, automotive control systems, sensor signal processing, uncertainty estimation, robust optimization methods, and probabilistic design.

Authors

Martin Cifrain studied Chemical Engineering at Graz University of Technology and received his PhD in 2001 in the field of Fuel Cells. From 2001 to 2003, he worked as lead engineer at Apollo Energy Systems in Florida and Germany, where he was responsible for the development of alkaline fuel cells. From 2003 to 2007, he was an independent consultant, mainly in the US. After a year working for EPCOS AG as an R&D engineer developing thermal sensors, he joined the VIRTUAL VEHICLE Research Center in 2008 as a senior researcher and project leader in several projects related to battery modelling. He is also a lecturer at the Joanneum University of Applied Sience.

David Fuchs is currently studying engineering physics at the Graz University of Technology. In 2010, he received a BSc in Technical Physics with an emphasis on

A. Thaler and D. Watzenig (eds.), *Automotive Battery Technology*,
Automotive Engineering: Simulation and Validation Methods,
DOI: 10.1007/978-3-319-02523-0, © The Author(s) 2014

materials science. He is currently working on his masters thesis, which focuses on thermal runaway experiments with consumer lithium-ion batteries.

Andrey W. Golubkov received his M.S. degree in engineering physics from the Graz University of Technology in 2008, after which he worked as a researcher at the Institute of Solid State Physics in the same university for 6 months. From 2009 to 2011, he was employed full time as a concept engineer at MAGNA STEYR Battery Systems GmbH & Co OG. Since 2011, on he has been working part time at MAGNA STEYR Battery Systems and part time as a researcher at the VIRTUAL VEHICLE Research Center. He is currently conducting his Ph.D. research on lithium-ion battery safety issues at the Graz University of Technology.

Heikki Haario is a professor of mathematics and the director of the Institute of Mathematics and Physics at Lappeenranta University of Technology in Finland. He is specialized in the modelling of industrial processes and inverse problems and the reliability analysis of models.

Christoph Hametner received an M.S. degree in mechanical engineering in 2005 and an PhD in 2007 from Vienna University of Technology. Currently, he is a research assistant at the Institute of Mechanics and Mechatronics, Vienna University of Technology. His theoretical areas of research are nonlinear system identification, modelling and control.

Stefan Jakubek received an M.S. degree in mechanical engineering in 1997 and a PhD in 2000. He completed his Habilitation (professorial qualification) in early 2007. From 2007 to 2009, he was head of development for Hybrid Powertrain Calibration and Battery Testing Technology at AVL-List GmbH. He is currently a Professor at the Institute of Mechanics and Mechatronics, Vienna University of Technology. His research interests include fault diagnosis, nonlinear system identification and simulation technology.

Werner Leitgeb is a senior researcher and project manager at the VIRTUAL VEHICLE Research Center in Graz. He received an M.S. degree in mechanical engineering and economics at Graz University of Technology in 2008. His main research areas are vehicle crash test and simulation methods including battery safety aspects.

Andrea Leitner is a senior researcher and project manager at the VIRTUAL VEHICLE Research Center in Graz. She received her Ph.D. degree in Information and Communication Engineering from Graz University of Technology in 2012. Her main research areas are methods and technologies for variability management, model-based software and systems engineering, tool interoperability and functional safety within the automotive product development process.

Helmut Martin received his M.S. degree in electrical engineering from Graz University of Technology in 2005. He worked for 6 years in the automotive industry, where he gained knowledge about model-based software development for an electric control system for a hybrid electric vehicle. He worked as a functional safety engineer for the development of a battery system for an electric vehicle (Functional

Safety according IEC 61508 related to ISO 26262). At the beginning of 2011, he joined the embedded system group and became a senior researcher and project coordinator at the VIRTUAL VEHICLE Research Center. His main research topics are the functional safety engineering for the automotive domain according ISO 26262 and embedded system development.

Sascha Nowak studied chemistry at Westfälische Wilhelms-Universität Münster, from which he received a PhD in analytical chemistry in 2009. He is currently the head of the competence area 'Analysis and Recycling' at MEET Battery Research Center.

Franz Pichler received his M.S. degree in numerical mathematics and modelling at the Karl Franzens University of Graz in 2011. Since then he is a junior researcher at the VIRTUAL VEHICLE Research Center in Graz. His main research areas are multi-scale modelling of batteries and the numerical simulation of such models.

Falko Schappacher studied chemistry at Westfälische Wilhelms-Universität Münster, from which he received a PhD in solid-state chemistry in 2008. He is currently the head of the competence area "Ageing and Safety of Lithium-Ion Batteries" at MEET Battery Research Center.

Matthias K. Scharrer studied Telematics at Graz University of Technology, from where he received his M.S. degree. He joined the VIRTUAL VEHICLE Research Center in 2010 and is currently a member of the battery research group. His research focus is on data processing, optimization and uncertainty estimation.

Gernot Trattnig studied physics and material science at the Graz University of Technology and the University of Leoben, where he received a PhD. He joined the Area Mechanics and Materials at the VIRTUAL VEHICLE Research Center in 2007. He is currently the team leader of the Materials & Forming Technologies group and works on the crash modelling of lithium-ion batteries.

Sascha Weber studied chemistry at Leibniz Universität Hannover and Westfälische Wilhelms-Universität Münster, from where he received his M.S. degree. He joined the research group of Martin Winter in 2009 and is currently a member of the competence area "Ageing and Safety of Lithium-Ion Batteries" at MEET Battery Research Center.

Bernhard Winkler received his M.S. degree in electrical engineering from Graz University of Technology. He worked in the automotive industry, especially in vehicle safety and testing. At the end of 2011, he joined the the embedded system group as a researcher at the VIRTUAL VEHICLE Research Center. His main research interest is the functional safety engineering for the automotive domain.